Henry How

The Mineralogy of Nova Scotia

A report to the provincial government

Henry How

The Mineralogy of Nova Scotia
A report to the provincial government

ISBN/EAN: 9783744790338

Printed in Europe, USA, Canada, Australia, Japan

Cover: Foto ©berggeist007 / pixelio.de

More available books at **www.hansebooks.com**

THE
MINERALOGY OF NOVA SCOTIA.

A REPORT

TO THE

PROVINCIAL GOVERNMENT.

BY

HENRY HOW, D.C.L.,

PROFESSOR OF CHEMISTRY AND NAT. HIST., UNIVERSITY OF KING'S COLLEGE, WINDSOR, N. S.;
CORRESPONDING MEMBER OF THE NATURAL HISTORY SOCIETY OF MONTREAL,
AND OF THE NEW YORK LYCEUM OF NATURAL HISTORY; FORMERLY
CHEMIST TO THE BRITISH ADMIRALTY STEAM
NAVY COAL ENQUIRY.

Halifax, N. S.:
CHARLES ANNAND, PUBLISHER, 11 PRINCE STREET.

1869.

PREFATORY LETTER.

KING'S COLLEGE, WINDSOR,
June 1st, 1868.

SIR,

I beg to lay before you the Report on the Mineralogy of the Province with the preparation of which I have been entrusted by the Provincial Government.

I have thought it proper to give in the form of an Introduction some account of the origin and scope of the Report, together with references to the numerous sources of official and other information on the minerals of the Province. The Table of Contents shews the heads under which the minerals are treated.

All which is respectfully submitted by,
Your obedient servant,
HENRY HOW, D. C. L.

The Honourable W. B. Vail,
Provincial Secretary.

CONTENTS.

CHAPTER I.
Introduction—References to Publications on the Minerals of the Province 1

CHAPTER II.
Coal and Allied Minerals 6

CHAPTER III.
Gold 37

CHAPTER IV.
Silver—Argentiferous Galena—Antimony—Mercury—Molybdenum—Arsenic—Cobalt—Nickel—Bismuth 58

CHAPTER V.
Copper—Ores of Copper—Ore of Lead—Ore of Zinc—Plumbago—Sulphur—Ore of Sulphur 65

CHAPTER VI.
Iron Ores—Mineral Paints 85

CHAPTER VII.
Ores of Manganese 110

CHAPTER VIII.
Gypsum—Anhydrite—Borates—Brine Springs—Salt—Magnesia Alum 127

CHAPTER IX.
Limestones—Marbles—Barytes—Moulding Sand—Clays 140

CHAPTER X.
Building Stones—Stones and Materials for Grinding and Polishing 169

CHAPTER XI.
Minerals for Jewellery and Ornamental Purposes not before mentioned....179

CHAPTER XII.
Minerals not included in the foregoing classes and chiefly adapted for the Cabinet..........186

CHAPTER XIII.
Mineral Waters..........192

CHAPTER XIV.
Catalogue of Localities of Minerals..........201

CHAPTER XV.
Notes on the Reservation of Minerals..........211

APPENDIX. (Page 216.)
Report on the Gold Regions, by Dr. S. Hunt—Gold Product of the Province—Copper Ore of Cape North, C. B.—Provincial Salt at the late Exhibition—Export of Minerals from Windsor in 1868—Marble in Cape Breton—Coal Mines.

ERRATA.

Page 23, and elsewhere, *for* Welch *read* Welsh.
,, 26, *last line* Ratio etc., *should be above first table.*
,, 43, line 16 from top *for* Sylwin *read* Selwyn.
,, 45, ,, 9 ,, ,, *for* Phillip's *read* Phillips's.
,, 52, ,, 3 ,, bottom, *for* Amalgamators *read* Amalgamaters.
,, 69, ,, 7 ,, top, *for* Cumberland *read* Colchester.
,, 103, ,, 2 from bottom, *for* it *read* titanium.
,, 121, ,, 18 ,, top, *for* Mushpratt *read* Muspratt.
,, 123, ,, 7 ,, bottom, *for* in *read* from.
,, 125, ,, 1 *for* veins *read* series.
,, 135, ,, 8 from top ⎫ selenite was shipped from Miller's Creek, Avon-
,, 137, ,, 2 ,, ,, ⎭ dale; price named was paid in New York.
,, 140, ,, 21 ,, ,, *for* Outram *read* Outrim.
,, 141, ,, 9 ,, ,, *for* ,, *read* ,,
,, 162, ,, 8 ,, ,, *for* with these results *read* the results were these.
,, 163, ,, 21 ,, ,, *for* fine *read* fire.
,, 164, ,, 10 ,, bottom, *for* fine *read* fire.
,, 173, ,, 7 ,, top, Tatamagouche should be in *Colchester County.*
,, 184, ,, 3 ,, ,, *for* simple *read* simply.

MINERALOGY OF NOVA SCOTIA.

CHAPTER I.

INTRODUCTION.

THE account of the Mineralogy of Nova Scotia now presented originated in the following way. The author, having soon after his arrival in the province in 1854 described some interesting minerals not before noticed here, was engaged in 1861 by the Provincial Commissioners of the Industrial Exhibition to be held the following year in London to make an illustrative collection of the minerals of the province for the Nova Scotian Court. The collection made included specimens from various museums and other sources, as noted in the official catalogue at the time, and it received the award of two medals, viz. of one in the Class of Mining, and of one in the Class of Educational Works and Appliances. It was the intention of those entrusted with the different divisions of the Nova Scotian Court to produce a joint report on the resources of the province, each contributor taking up his particular department. The minerals fell to the author, and he prepared his account of them. The plan, however, of such a report which no doubt would have been very useful, was not carried out, and the author subsequently from time to time sent papers to the Nova Scotia Institute, entitled Notes on the Economic Mineralogy of Nova Scotia, consisting of portions of his manuscript, with necessary additions. These papers were published in the Transactions of the Institute, and were said to be of interest, and of some use in drawing attention to the minerals described. Much information had been given on the minerals of the province in Dr. Dawson's admirable Acadian Geology, and the papers in question made frequent reference to his pages, while they went more fully into certain details than would have been consistent with the objects of the "Geology." The author prepared for the Provincial Commissioners a second collection of minerals for the Dublin Exhibition, in 1865, which received the award of a medal, and a third for the late Paris Exhibition, for which Honorable Mention was awarded. On this last occasion the author wrote a "Sketch of the Mineralogy of Nova Scotia, as illustrated by the specimens sent to the Paris Exhibition," for the official catalogue. This sketch, of which separate copies were issued by authority of the Provincial Commissioners, was reported by Dr. Honeyman to have been very useful in Paris, and it was thought here that it might be advantageously expanded so as to include as much as necessary and convenient of the information in existence relating to the minerals of the province. The matter having been, after some conversation, formally laid before the late Government, it was resolved that the author's proposal to write on the minerals of the province should be accepted. The report now presented is in fulfilment of the undertaking in question.

The object of the author is to give such a general view of the mineral resources of the province as will shew what has actually been done in the working of its best known treasures, coal, gold, iron, gypsum and building stones, and at the same time draw attention to other minerals of considerable importance, the uses of which, as well as what can be ascertained of their money value, are pointed out. Diligent search having been made for statistics as to the quantity of minerals produced, and for the results of chemical analysis of the various kinds of minerals, a good deal of information has been collected on these points, and, where practicable, thrown into the form of tables which will probably be found among the most interesting features of the report. It has been thought desirable, in order to have a complete account of the mineralogy of the province, not to omit a brief statement of the most important facts respecting those minerals which are not of economic value, but whose beauty, rarity, or scientific relations render them of great interest; and also such mineral waters as have been examined or are known or supposed to possess medicinal properties.

The sources of special information given for the sake of comparison and illustration are indicated in the body of the report, and it is not necessary to give a list of the numerous books and periodicals referred to on those points, but it will be very useful to mention the principal publications on the minerals of the province; the majority of these have been consulted on the present occasion.

The Journals of the House of Assembly contain a large number of documents on Mines and Minerals. The arrangement between the Provincial Government and the General Mining Association of London, by which the mines of the province were, except in limited tracts, thrown open to general enterprise, can be traced in all its stages up to its completion in 1858. After this year come numerous documents on the gold fields, and reports of the Inspectors of Mines, of the Chief Gold Commissioners, and of the Chief Commissioners of Mines. Of reports on Geological Surveys there are the following :—

On Coal Fields of Carriboo and R. Inhabitants, J. W. Dawson..1846.
On Discovery of Gold in N. S., Joseph Howe and H. How..1860.
On Gold Fields, Joseph Howe.................................1861.
On Gold Fields, Western Section, Henry Poole........ } 1862.
Analysis of some of the Minerals collected, Henry How. }
On Gold Fields, Eastern Section, John Campbell...........1862.
On Gold Fields, John Campbell...........................1863.
Geo. Survey of Nova Scotia and Cape Breton, D. Honeyman } 1864.
Appendix on Minerals, H. How........................ }
On Minerals collected by D. Honeyman, 1864, H. How.....1865.

The general mineralogy of the province is treated more or less fully in :—

Remarks on the Mineralogy and Geology of N. S., C. T. }
   Jackson and F. Alger:............................. } 1833.
Memoirs of American Academy ; Vol. 1................. }
Remarks on Geology and Mineralogy of N. S., A. Gesner....1836.
Acadian Geology, J. W. Dawson..........................1855.
Supplementary Chapter " ....1860.
Sketch of the Mineralogy of N. S., H. How..............1867.
Acadian Geology, Second Edition, J. W. Dawson..........1868.

The following treat on special subjects :—
The Gold Fields of N. S., A. Heatherington..............1867.

The Transactions of the N. S. Institute contain these **papers** connected with the minerals of the province:—

On Magnesia Alum or Pickeringite, H. How.................1863.
On Mineral Springs of Wilmot, " ....1864.
On Rocks in the vicinity of Halifax, W. Gossip............1864.
Notes on the Economic Mineralogy of N. S., Part I, } 1864.
  Iron Ores....................................H. How. }
On Brine Springs of N. S.........................." ...1865.
Notes on Ec. Min. N. S., II, Ores of Manganese. " ...1865.
On Auriferous Deposits of N.S., P. S. Hamilton, Ch. Gold Com..1866.
On some recent Improvements in the Amalgamation } 1866.
  Process for extracting Gold from Quartz....G. Lawson }
Notes on Ec. Min. N. S., III, Limestone and Marble, H. How 1866.
Remarks on Minerals sent to Paris Exhibition..... " 1867.
Geology of Gay's River Gold Field, D. Honeyman..........1867.
Explorations in Pictou Coal Field, R. G. Haliburton.........1867.
Geol. Features of Londonderry Iron Mine, D. Honeyman....1867.
Notes on Ec. Min. N. S., IV, Gypsum and Anhydrite, and } 1868.
  the Borates and other Minerals they contain,..H. How, }

The following reports to companies made here have been consulted:—

On the Lead Ore of Gay's River, F. Bawden, Johnson } 1862.
  and Matthey and others......................... }
On Manganese Ores of Tenny Cape, H. How.................1864.
" " " and Cape Burton, D. Honeyman.........1865.
On East Mountain of Onslow Lime and Manganese Co's. } 1866.
  Property, W. Barnes, D. Honeyman, H. How........ }
On an Iron Deposit at Brookfield, W. Barnes, H. How......1866.
On Cement and Paint Stone of Chester Basin, H. How......1866.
On Gold Field of Gold River................." ......1866.
On Montreal and Pictou Co's, Coal, etc........" .....1866.
On the Silver Mine, etc., at Baddeck, H. Y. Hind..........1867.
On the Baryta of Five Islands, G. D. Whycoff..............1867.
On the Douglas Slate Quarry, G. Lawson...................1868.
Reference has been made to the successive numbers of } 1868.
  the "Mining Gazette," Halifax, A. Heatherington... }

The following papers on special points of N. S. mineralogy have been published abroad:—

On Natroborocalcite in Gypsum of N.S., H. How, Silli- } 1857.
  man's Journal.................................... }

INTRODUCTION.

Analysis of Faröclite and some other N. S. Minerals, H. How, Silliman's Journal........................ } 1858.

On Three New Minerals from Trap of N.S., H. How, Edin. New Phil. Jnl.................................. } 1859.

On an Oil Coal from Pictou Co., N. S., H. How, Sill. Jnl. and Ed. N. P. Jnl................................ } 1860.

On Gyrolite in Trap of N. S., H. How, Ed. N P. Jnl. and Sill. Jnl................................... } 1861.

On Natroborocalcite and another Borate in Gypsum of N. S., H. How, Sill. Jnl........................ } 1861.

On the Gold of N. S., O. C. Marsh, Sill. Jnl............1861.

On the Gold Fields of N. S., A. Gesner, Geol. Soc. London..1862.

On Pickeringite in N. S., H. How, Jnl. Chemical Society, London............................... } 1863.

Report on N. S. and N. Y. Co's. Gold Property, B. Silliman..1864.

On Mordenite, a new Mineral from Trap of N. S., H. How, Qu. Jnl. Chem. Soc., London............... } 1864.

On Gold Mines and Gold Mining in N. S., H. Perley, Can. Naturalist.................................. } 1865.

Contributions to the Mineralogy of N. S., Part I, Manganite, Pyrolusite, Wad, H. How, London Edinburgh and Dublin Phil. Mag............................ } 1866.

Do. Part II, Wichtisite, Pencil Stone, Variegated Slate, Bitumen in Calcite, H. How, L. E. D. Philosophical Magazine... } 1867

On Ledererite in N. S., O. C. Marsh, Silliman's Jnl.........1867.

Some account of the operations of Gold Mining in N. S., G. Lawson, Chemical News...................... } 1867.

Contributions to Mineralogy of N. S., III, Silicoborocalcite, a new Mineral, etc., and Addendum, H. How, L. E. D. Phil. Mag................................. } 1868.

There has also been published :—

A Catalogue of Mineral Localities in New Brunswick, Nova Scotia, and Newfoundland, O. C. Marsh, Silliman's Jnl. } 1863.

The author is indebted to several gentlemen for private communications, as specially mentioned in the body of the report, and he has included numerous unpublished notes of his own. For the valuable information contained in the chapter on the reservation of minerals in the province, he is obliged to W. A. Hendry, Esq., Deputy Commissioner of Crown Lands.

CHAPTER II.

COAL AND ALLIED MINERALS.

COAL is the most important of the minerals of the province, it was worked in Cape Breton by the French, upwards of a century ago, and no doubt it has been raised ever since to some extent. During the last forty years, though for the first thirty mined under a practical monopoly, it has been produced to the ascertained value of about 20 millions of dollars. In the year 1826, a grant was made by George IV., to the Duke of York, of all mines and minerals in the province, not previously granted with the land, for the term of 60 years, at certain rents and royalties. The lease was assigned to some creditors of the Duke, and transferred to an English company, called the General Mining Association, who commenced operations here in 1827. After the Provincial Government had made prolonged efforts to remedy a state of things so manifestly prejudicial to the interests of Nova Scotia, an arrangement was come to in 1857, and the next year an Act was passed, by which the rights of the Association were restricted to the mining of coal alone, in six areas, selected by themselves. This proceeding at once gave life to mining enterprise, and as regards coal the result has been that while the Association worked only six mines in two counties of Nova Scotia proper and one county of Cape Breton, returns of coal have since been made from 36 mines in two counties of Nova Scotia, and the four counties of Cape Breton, seams from two to nine feet thick have lately been reported in a third county of Nova Scotia, and several new companies have just been organized for the working of valuable properties. The coal mines were from 1858 to 1860 under the supervision of a Provincial Inspector of Mines, J. McKeagney, Esq., who made most valuable reports upon the subject, giving the nature and extent of the operations of the

Association and the amount of coal sold from each of their mines, and the results of the action above mentioned in the opening of new mines, and the raising an increasing quantity of coal, with much further important information. In 1861 the office of Inspector of Mines was abolished, and the duty of inspection transferred to the Commissioner of Crown Lands, by whom detailed reports were issued till 1864, when the greatly and rapidly increasing business of mining rendered other arrangements necessary. The Department of Mines was then formed, and placed under the control of the Chief Commissioner of Mines, and the office of Inspector of Mines was re-established. In this latter capacity, John Rutherford, Esq., M. E., Member of the North of England Institute of Mining Engineers, made his first report in 1866. In this and his report for the following year full particulars as to the actual condition of all the works are given, while the reports of the Chief Commissioners of Mines with which they are issued afford a continuation of details respecting the amount of coal raised and sold and other matters of importance. To an abstract of some of the most essential points given in all these voluminous reports and in other sources will be added the principal results of such examinations into the qualities of the coals by analysis and practical trials, as I have been able to collect or have myself made.

The productive coal measures of the province so far as known (I am not aware of the exact age of the beds lately reported in Antigonish county,) are the second in descending order of the five members into which Dr. Dawson divides the carboniferous system. They are found in the counties of Pictou, Cumberland, Hants, and Colchester, Nova Scotia, and in the four counties of Cape Breton, covering an area the known extent of which is very large. Thus from late official reports it appears that in 1865 there were in addition to the territory of the General Mining Association (comprising 44 square miles) 31 square miles under coal mining leases. The aggregate extent of areas under licenses to search was about 1920 square miles. Within the year 1866, 376 applications were made for licenses to search, embracing about 1880 square miles, of which 84, covering about 420 square miles, were for ground never previously applied for. The number of licenses to work taken out the same year comprised 73 square miles, a larger tract than had ever been applied for within any previous year. This last fact is mentioned as indicating an increased degree of confidence in those who have been most engaged in explorations. In Cape Breton county alone, according to Mr. R. Brown, formerly Agent at the Sydney Mines, the productive coal measures cover 250 square miles, their thickness being probably 10,000 feet.

The following extracts from the official reports shew the late progress and present condition of coal mining. In 1858, the first year of the working of the New Mines under the Act before mentioned, the quantity of coal sold from the mines of the Association, viz.: those at Albion Mines, Joggins, Sydney, Lingan, and Aconi, was 224,400 tons, of 2240 lb. and from the New Mines, 8 in number, only 2,325 tons; in 1865, the most prosperous year recorded, the Association sold 366,962 tons, while the New Mines, 22 in number, sold no less than 285,892 tons. In this last year the general results of coal mining had been no less satisfactory than those of gold mining; thirty collieries were in operation, some of them, only just opened, had made but small returns, but in all, with one or two exceptions, works were being vigorously prosecuted with good prospects; the returns shewed the total quantity of coal sold during the year, ending 30th Sept., to be 652,854 tons. In 1866 the total sale of coal was 601,302 tons; the decrease was owing to the abrogation of the Reciprocity Treaty made with the United States in 1854 and the imposition in the latter country of a somewhat heavy duty, $1.50 per ton, on coal; the effect of these measures on the coal trade was of course damaging, as the United States was the largest consumer. Still, the effect was not so great as might reasonably have been expected, and the aspect of affairs at the close of the first fiscal year after the abrogation of the Treaty was the reverse of discouraging. While the total decrease in the sale of coal, as compared with the previous year, was 51,552 tons, the shipments to the United States shewed a decrease of 145,744 tons. This falling off was considered to be not wholly due to the abrogation of the Treaty. The great demand for coal during the late war, and the depressing effects of that war upon productive industry in the United States, gave a great stimulus to the coal trade here, and one which did not cease with the close of the war. Again, when the abrogation of the Treaty was imminent, a further stimulus was afforded, efforts being made to force as much coal as possible into the United States before the imposition of a duty.

Prospects were very cheering in the direction in which the coal trade had increased. The proprietors of collieries having received a check from the United States had looked round for new markets. The home consumption had increased of course; the actual increase being about 50 per cent, within the year. What was more important was the fact of the exports to the neighbouring colonies having increased by 54,099 tons. These figures, however, did not sufficiently explain the matter. The annual export of coal to the neighbouring colonies had more than doubled within the year and the indications warranted belief in a rapid and continued increase in the trade. In 1867 the total deficiency in the coal sold as compared with 1866 was 119,224 tons which was attributed to the abrogation of the Reciprocity Treaty. There was reason to hope that the existing arrangements with the United States would be modified. The expectations that the loss of the trade with the latter country would have been made up by increased home consumption and exports to neighbouring colonies were not realized, though the falling off was chiefly in "other countries." There was a decrease in home consumption of 1,983 tons, and in exports to neighbouring colonies of 3,379 tons. Although the returns of coal raised shewed a decrease of nearly 21 per cent. and work had been suspended at some of the new mines, it was a cheering fact that not only had additional mines been opened, but preparations were being made at others for a considerable extension of the powers of production. As regards the capabilities for supplying an extensive demand, Mr. Haliburton, in an article in the Transactions of the Nova Scotia Institute, says: Nova Scotian collieries now opened or in preparation would raise, in 5 years, 5 or 6 million tons annually, and the supply could be gradually increased to meet any demand, however great.

Frequent illustration has been made in foreign Exhibitions of the coal of the province. Mr. Poole, Agent to the General Mining Association, sent to New York in 1853, a continuous specimen of the whole Main Seam at the Albion Mines, Pictou county, shewing a thickness of 38 feet 6 inches. Mr. Scott, Agent, sent a similar section to Montreal in 1855, and a column to London, in 1862, when a medal was awarded. A column was sent to Dublin, in 1865, for which a medal was given, and which remains there in the Winter Garden, with its name, origin, height, 35 feet 6 inches, as well as the honors obtained in 1862 and 1865 properly indicated. In Paris, in 1867, for a column 37 feet 10 inches high, sent by Mr. J. Hudson, Agent, a Silver Medal was awarded to the General Mining Association. At Dublin, in 1865, the coal of Cape Breton was distinguished by the award of four Honourable Mentions, viz.: to Hon. T. D. Archibald, for " good sample," to Mr. R. Brown, for "interesting specimens," to Mr. J. C. Campbell, for "a good specimen," and to Messrs. Symonds, Kay, and Ross, for " a good specimen " of coal. At the late Paris Exhibition, in addition to the Albion column, there were shewn :—

A column of Coal 5 feet thick, from Gowrie Mines, Cape Breton,
" " 5 " " " Sydney Mines, " "
" " 8 " " " Caledonia Mine, " "
" " 9 " " " Cow Bay Mine, " "
" " 9½ " " " Little Glace Bay Mine, C. B.

And Oil Coal and Oil from Albion Coal Fields.

The following table shews some of the leading facts connected with the coal mines of the province, from which coal has been raised of late years, they are 36 in number :—

MINE.	PRINCIPAL SEAMS.			WORKINGS.		Coals raised in tons of 2240 lbs.	
	No.	Thickness.	Dip.	Greatest Depth.	Extent.	1866.	1867.
Cumber'd Co., N.S Joggins............	2	6 ft. 2 in. / 3 " 4 "	S.W. 19° or 1 in 2.9	Shaft 110 ft.	25 acres.	8,478	8,806
Victoria	1	5 " 1½ "	S. 17° or 1 in 3.5	Shaft 135 "	....	2,077	290
Lawrence	2	2 " 4½ "	S. 22° or 1 in 2.5	Slope	6	....	....
Macan............	1	2 " 4½ "	S. 35° or 1 in 1.5	Slope 150 "	....	2,300	830
Chiegnecto......	1	12 " 9 "	S. 42° or 1 in 1.2	Shaft 90 "	....	4,847	212½
St. George......	1	11 " 9½ "	S. 46° or 1 in 1	Slope 210 "	....	150	....
Pictou Co., N. S. Albion	2	38 " 6 " / 15 " 6 "	N.E. 20° or 1 in 2.75 / N.E. 20° or 1 in 2.75	Shaft 840 " / Shaft	100 " / 60 "	222,437	143,334
Acadia	2	12 " 0 " / 20 " 0 "	N.E. 20° or 1 in 2.75 / N.E. 20° or 1 in 2.75	Slope 550 " / Slope 390 "		14,662	18,726
Nova Scotia....	1	19 " 10 "	E. 20° or 1 in 2.75	Slope 225 "	....	62	41
Bear Creek.....	1	19 " 10 "	E. 20° or 1 in 2.75	Shaft	....	533	443
McKay............	....	....	....	....	....	176	281
Pictou (German)	....	....	....	....	....	....	164
Mont. & Pictou.	1	25 " 0 "	S. E. 65°	Shaft 180 "	....	....	421
M'Bean...........	....	....	....	....	....	....	22
Inverness Co, C.B Port Hood......	1	6 " 1 "	N. W. 27° or 1 in 2	Slope 300 "	....	3,824	6,315
Chimney Corner	....	....	....	....	....	....	50
Victoria Co., C.B. N. Campbeltown	3	6 " 0 " / 4 " 0 " / 4 " 5 "	E. 12° or 1 in 5	Slope 300 "	....	3,142	4,033
Black Rock.....	....	....	....	....	....	....	209
Cape Breton Co. Matheson, Lit. Bras d'Or..,	1	2 "11½"	E. 8° or 1 in 7	....	....	468	775½
Collins	1	4 "11¾"	E. 6° or 1 in 10	Shaft 90 "	....	....	433

(Continued.)

MINE.	PRINCIPAL SEAMS.			WORKINGS.		Coals raised in tons of 2240 lbs.	
	No.	Thickness.	Dip.	Greatest Depth.	Extent.	1866.	1867.
Cape Breton Co.							
Sydney	2	6" 0" / 6" 3½"	E. 7° or 1 in 8	Shaft 390 ft. / Drift 3316 "	400 acr's	132,915	116,533
Ingraham	1	4" 0"				10	53
Lingan	1	5" 5"	N. E. 12° or 1 in 4.7	Slope 600 "	45 "	59,780	45,626½
International	1	5" 6"	E. 5° or 1 in 11	Slope 300 "	7 "	13,364	15,957½
Caledonia	1	8" 3"		Shaft 173 "		587	8,015
Little Glace Bay	2	9" 9½" / 5" 0½"	E. 5° or 1 in 11	Slope 400 " / Shaft 111 " / Shaft 36 "	25 "	61,902	46,716½
Acadia						374	......
Clyde	1	8" 6"	N. E. 7° or 1 in 8	Slope 480 "		7,153	20
Schooner Pond	1	6" 11½"	N. 6° or 1 in 3.7	Slope 240 "		(6,206)	......
Block House	1	8" 10"	N. E. 5° or 1 in 11	Slope	12 "	107,642	84,938
Gowrie	1	4" 11"	N. E. 7° or 1 in 8	Shaft 80 "		35,704	38,532
Mira Bay	1	4" 6"	N. E.	Slope 270 "		(2,893)	......
South Head	1	3" 6"	E.	Shaft 71 "		1,138	......
Victoria							350
Richmond Co, C. B							
Richmond	2	3" 0" / 4" 0"	N. E. 85°	Shaft 200 "		1,016	......
Sea Coal Bay	....	4 to 7 ft.				(400)	......
				Total		684,740	542,127½

A larger quantity of coal was raised in 1865, namely, 712,575 tons, the greatest annual weight yet produced. The years following are selected, as shewing the greatest number of mines in operation; the amounts in brackets were got out in 1864 or 1865, and are inserted to shew that the mines have been worked, they are not included in the total results.

Many of the new mines, in Cape Breton and Cumberland counties especially, have been opened on seams long known to exist. In Pictou county there has been a most remarkable development of the productive beds. Mr. Rutherford says of those large seams worked at the Acadia, Nova Scotia, and Bear Creek mines, about 20 feet in thickness, and others underlying, so far (three miles to the west, for example, and again to the south, at varying distances,) from the series proved by the General Mining Association and Acadia Company, that whatever may be the case as to their identity with those long known, it is a gratifying and important fact that seams of coal of an exceedingly valuable character have been traced over a tract of country in which their existence was only a short time ago extremely problematical. To what extent they may spread existing openings scarcely afford a sufficient basis for conjecture; but considering their inland position and consequent freedom from the limit of yield to which the subaqueous coal fields are subject, their economical importance cannot be too highly estimated. With reference to the great discoveries made to the north and east of the Albion Mines, the coal found by Mr. Haliburton in several beds of from 2 feet 6 inches to 25 feet in thickness on the Montreal and Pictou area is thus spoken of: "The discovery of coal on this area has added to the importance of the Pictou coal field in a remarkable degree;" and it is said also, "The same seam has been discovered by Mr. Kirby on the east side of East River, to the north of New Glasgow. This extension of this portion of the coal field will doubtless lead to further explorations, the progress of which will be watched with interest." Prospectings were made in several places in 1866 with more or less success. Mr. Kirby proved two seams of coal, each four feet thick, about two miles east of New Glasgow. A seam of coal was found by Mr. Haliburton, at Sutherland's river, about seven miles east of Albion Mines, it proved to be 14 feet total thickness at 100 feet depth, but disturbed, and **further** explorations **were in** progress; between this and the Albion Mines on the M'Bean area, coal was opened upon. From these and other details it is evident that the limits of the coal basin here have been largely extended in every direction. **In connection** with these operations notice is made of the **recently reported discovery of coal near** Antigonish, in the county of that name, by Messrs. McKinnon and Chisholm. This coal field, it is remarked, is an interesting addition to those already known, and its development will be attended with much interest. It is separated from the county of Pictou by metamorphic rocks.

Cumberland county is spoken of in the report of 1864, as having a large tract of country of which Spring Hill may be taken as the centre which has long been known to contain exceedingly rich beds **of coal.** Explorations carried on with much spirit had been attended with cheering results. Under ordinary conditions mines could not be opened and successfully worked in this inland district without large preparatory expenditure. But if, as is anticipated, the projected Intercolonial Railway should pass through, or skirt this coal basin, a number of rich coal mines will, beyond question, be opened.

In Inverness county, Cape Breton, explorations have proved the existence of several seams of coal along the coast to the east of

Port Hood. At Mabou, Broad Cove, and Chimney Corner, seams varying in thickness from 3 to 7 feet, have been found.

At the close of his first report, in 1866, Mr. Rutherford makes some remarks, an abstract of which will give a very good general idea of the character of the coal mines. The remarks are said to apply with a few exceptions to nearly all the collieries of the province. It is mentioned that the facility with which coal has been reached in all the districts, as compared with other mining countries in which, from the exhaustion of the seams near the crop, expensive sinkings become necessary, instead of permitting an effective winning of a large tract of coal before beginning the regular working, seems to have engendered an indifference to future operations, and allowed the desire for immediate profit to supersede the necessity of a judicious arrangement of the mode of working. To this cause is attributed the short distance from the crop to which, in most of the mines, the workings are confined. The introduction into the market of coal worked so close to the crop, is believed to have operated injuriously to the interests of the mine owners. The small size of the pillars left is objected to, for reasons given in detail. The general freedom from gas is mentioned as having produced an indifference as to the necessity of making provision for effective ventilation as the workings become extended. The mere fact of there being at present in most of the mines no deleterious gases, should not be deemed sufficient to render unnecessary the provision of the fresh air requisite to the healthy existence of the miners. In many instances the exposure of the coal on the faces of cliffs has induced the opening of the seams to be made by driving an adit or level from the shore, which has answered the double purpose of being an outlet for the coal and the water made in the mine. The desire to obtain as much coal as possible has in many cases led to this level being so placed that it is within reach of the tide, which occasionally flows into the mine. Dams have been fixed in some of the collieries to prevent any unusual rise of the tide overflowing the dip workings, and their adoption in others is recommended. The quantity of water made in the mines is not large, and it is thought that much of it is from the surface. The great loss consequent upon storing the coal got out in winter on the surface is pointed out. The production of slack coal by exposure and by breaking in putting down and relifting is such that it is believed in some instances not much more than fifty per cent. of large coal is obtained from the heap. Suggestions as to the means of lessening this great loss by modification of the plan of working the mines and erection of suitable protective buildings on the surface are made.

The total amount of coal **raised and shipped** in the province is given in **the** Inspector's report for 1863 as being, from 1827 to 1857, inclusive, 3,692,762 tons. From the year the new mines were opened (1858) the returns of coal up to last year I **have** calculated from official data to be as follows :—

COAL RAISED AND SOLD IN NOVA SCOTIA IN TONS OF 2240, LBS.

Year.	Total Raised	Raised and Sold.	Raised and sold from New Mines only.	
1858		229,950 tons.	2,325 tons.	
1859		270,293 "	8,908 "	inc. 2151 tons Oil coal.
1860		322,302 "	13,352 "	" 1643 " " "
1861		326,428 "	41,530 "	
1862		394,708 "	41,614 "	
1863		429,351 "	85,959 "	
1864, (9 mos)	395,030 tons.	406,699 "	152,358 "	(raised 151,715 tons.)
1865	712,575 "	652,854 "	285,892 "	" 292,789 "
1866	684,740 "	601,302 "	201,586 "	" 261,130 "
1867	542,127 "	482,078 "	217,378 "	" 227,828 "
From 1858 to 1867......		4,115,965 tons.	1,050,920	
From 1827 to 1857......		3,692,762 "		
Total raised and sold..		7,708,727 tons.		

The coal is sold at the collieries at prices varying from $1.70 to $2.50 per ton for large, and 80 cents to $1.20 for small. The average price for large coal is $2 per ton at the New Mines. At Little Glace Bay it is put on board for that **sum.** At Lingan the price is about ten cents higher, and at **Sydney** and the Albion Mines, the highest price named, $2.50, is obtained. It was one of **the terms of the** agreement before mentioned **as** entered into between the Provincial Government and the General Mining Association, that during the lease held by the latter, the royalty on all coal raised in the province, should be 10 cents per ton of 2240 lbs. on all quantities up to 25,000 tons raised by each company, with one third diminution on all over that quantity ; small, or slack coal, and coal used by the workmen, and in carrying on the works, to be free of royalty ; it was also stipulated that during the lease no export duty should be placed on coal without the consent of the Association. The royalty accruing to the province I find to be as follows, for those years in which returns of amount of coal raised are given:—

	Royalty paid by Coal.	Actual Royalty on Coal raised.
1863	$30,959.	
1864 (9 mo.)	33,745.	$39,503.
1865	43,645.	65,597.
1866	46,939.	62,687.
1867	64,486.	52,068.

On this subject the report of 1867 states that the increase observed arises from two causes. In 1865, an amendment to the **Mines** Act was passed, making the royalty payable quarterly, whilst previous to that date, it was payable yearly, so that part of the income on coal raised had been paid during the current year. There had also been some progress made in collecting arrearages, of which there is still a considerable amount due from former years, as shewn in the statement given.

Quality of the coals.—The coal of the province is bituminous, or soft; no anthracite or hard coal has been met with. There are also very rich oil-coals, distinguished from common coals by their composition and property of yielding oil on careful distillation, and more resembling cannel coal, which is also found here, but not abundantly. In giving the results of examination into the qualities of the coals, by chemical analysis and practical trials, I shall take them by counties. I must mention that some of the analyses were made several years ago, and probably the character of some of the coals would now be found different; as a general rule, improvement of quality might be expected in increased depth of workings. As regards the "evaporative power" I must state that when it is mentioned as "theoretical," as it is in most cases, it is calculated from a formula employed in the British Admiralty Coal Enquiry on coals suited to the Steam Navy, in which I was formerly engaged as chemist, and expresses the number of pounds of water which the coke by itself could convert into steam, at the temperature of $212°$ Fahrenheit. It was found in the Enquiry that this theoretical result, notwithstanding several striking exceptions, showed that the work capable of being performed by the coke alone, is actually greater than that obtained under the boiler with the original coal. In the first report issued on the subject, it appeared that out of 29 coals examined, the theoretical number was greater than the practical in 16 cases, the reverse being found with the remainder. In the absence of practical trials, therefore, this mode of representing the value of a coal for steam purposes, can only be taken as approximatively correct.

Coal of Cumberland County.—The coal from the Joggins was described by Dr. Dawson, in 1855, as free-burning bituminous coal, of fair quality, affording, on analysis of a specimen from the main seam,—

Moisture 2.50
Volatile Combustible Matter............36.30
Fixed Carbon..........................56.00
Reddish-gray Ashes..................... 5.20
                                        ——————
                                        100.00

The specimen was bright coal, of uniform texture, with straight joints, containing films of iron pyrites and calcareous matter. The principal market for this coal is St. John, New Brunswick.

The coal of Springhill, (South) about 20 miles south east of Joggins shore, has not been worked for exportation, as, owing to its inland situation, it would not at present be remunerative. It is expected, however, that the opening of the projected Intercolonial Railway will make this deposit very valuable. The coal, of which one seam is twelve feet thick, dipping north, has been examined by Dr. Dawson and myself, with the following results:—

	Dawson.	How.
Moisture	1.80	2.92
Volatile Combustible Matter.......	28.40	22.46
Fixed Carbon..................	56.60	60.95
Reddish Ashes	13.20	13.67
	100.00	100.00
Theoretical evaporative power............		8.37

Dr. Dawson states that from the character given of this coal by persons who have used it, he should infer either that the quality had been overrated, or that his specimen was inferior to the average quality. The latter is probably the case, as the following analysis by Messrs. Woodhouse and Jeffcocke, of Derby, England, for which I am indebted to E. A. Jones, Esq., of the Acadia Mines, shews the coal to be of a quality superior to that indicated above.

Analysis of coal from 12 feet seam, Springhill:—

Woodhouse and Jeffcocke.

Carbon	72.00
Hydrogen	5.02
Oxygen	7.26
Nitrogen	1.96
Sulphur	0.84
Water	2.60
Ash	10.80
	100.56

Mr. Jones remarks that the Intercolonial Railway will bring the Acadia Iron Mines into communication with the Springhill coal fields, distant 24 miles, and will thus assist very much in developing the iron interest.

The North Spring Hill seams are between five and six miles from the River Napan, and dip south. They have been proved to be three in number but do not appear to have been actually worked, operations were begun in 1859 and soon discontinued. In the Inspector's report of the next year it is said that the coal is pronounced to be of superior quality. The following analysis is given :—

Volatile matter	37.000
Fixed carbon	59.174
Ash	3.826
	100.000

Sulphur in volatile matter 0.316.

It is said that "the almost entire absence of sulphur in the volatile matter renders the coal especially adapted for gas purposes."

The Victoria coal is much esteemed in St. John, New Brunswick, where it is chiefly sold. The Macan (Lawson's) coal has been found a good domestic coal at Windsor where it has sold at $4.50 per chaldron. The only objection made to it is that it forms clinkers in stoves.

Coal of Pictou County. The coal from the Albion Mines was very thoroughly tested in 1842 or 1843 by Professor Johnson in the experiments made upon coals at the Washington Navy Yard by order of the United States Government. The composition of the two samples of coal used was as follows:

	I.	II.
Moisture	2.567	0.781
Volatile combustible matter	27.063	25.795
Fixed carbon	56.981	60.735
Ash	13.389	12.508
	100.000	100.000
Sulphur		0.769
Specific gravity	1.320	1.330
Weight of a cubic foot in a merchantable state	53.55 lb.	49.25 lb.
" " " the mine, calculated, about	82½ lb.	
Actual evaporative power, viz: pounds of water evaporated at 212° by one pound of coal	8.41	8.49
Theoretical evaporative power (I have calculated to be)	7.82	8.34

No. 1 was easily ignited, its clinker was in sheets of considerable magnitude and somewhat porous. No. 2 burnt promptly with a long smoky flame, its clinker was black, vitreous and porous, tolerably friable and not apparently inclined to adhere to the grate.

The evaporative power of bituminous coals from Liverpool, and Newcastle, England, and a locality in Scotland was found *under the same circumstances* to be respectively 7.84; 8.66; and 6.95.

It will be useful to state here that the results of the British Admiralty Coal Enquiry were these:—

No. and locality of British Coals examined.	Actual evaporative power, or No. of pounds of water evaporated from 212° by 1 pound of coal.
Average of 37 Coals from Wales	9.05
" " 17 " Newcastle	8.37
" " 28 " Lancashire	7.94
" " 8 " Scotland	7.70
" " 8 " Derbyshire	7.58

Numerous analyses were made by Dr. Dawson in 1854 shewing the character of the Albion Mines coal from different parts of the upper floor of the mine, and also the varieties existing throughout the whole thickness of their main seam, in a series of assays of coal taken at distances of one foot in thickness. The general results were that the best coal was found on the N. W. side of the old workings, deterioration taking place at either extremity of the workings of the upper floor. In all parts of the mine the lower coal was inferior to that of the middle of the seam and still more so to that of the upper part above the "holing stone" or "fall coal" of the miners. On the west this fall coal disappeared **or was** reduced to insignificant thickness. The assays made to show the variations in thickness of the whole seam were on coal taken at this western part. This valuable series of assays of the coal of **this seam so** familiar to the world is here given.

Assays of Samples taken at distances of one foot in thickness in the Main Seam of Coal at the Albion Mines, Pictou, by Dr. Dawson:

	Volatile matter by rapid coking.	Volatile matter by slow coking.	Fixed Carbon.	Ashes.
No. 1. Coal	26.0	19.9	68.8	16.3
2. do.	27.8	24.1	63.8	12.1
3. do.	27.4	25.7	60.0	14.3
4. do.	27.2	25.0	65.5	9.5
5. do.	25.8	25.1	64.8	10.1
6. do.	25.2	24.9	62.5	12.6
7. do.	27.4	22.0	68.5	9.5
8. do.	26.8	22.9	66.7	10.4
9. do.	27.0	23.9	61.3	14.8
10. Carbonaceous shale	16.4	15.9	26.3	58.8
11. Coal	28.8	25.8	59.7	14.5
12. do.	27.2	25.4	62.5	12.1
13. do.	27.6	24.7	62.5	9.8
14. do.	26.6	23.9	61.0	15.1
15. do.	26.8	23.1	65.1	11.8
16. do.	28.8	24.9	62.3	12.8
17. do.	30.4	26.0	65.0	9.0
18. do.	26.0	26.1	63.0	10.9
19. do.	26.0	25.0	66.3	18.7
20. do.	26.8	22.7	63.6	13.7
21. Coarse Coal	25.8	23.3	58.3	18.4
22. do.	27.2	22.5	60.3	17.2
23. Coal	29.4	23.6	64.3	12.1
24. Coarse Coal	25.8	22.4	57.6	20.0
25. do.	25.8	23.1	60.2	16.7
26. do.	27.8	21.9	54.8	23.3
27. Coal	27.0	24.3	65.5	10.2
28. do.	25.6	22.4	65.0	12.6
29. do.	25.8	22.7	62.7	14.6
30. do.	27.2	23.1	67.4	9.5
31. do.	32.6	22.4	66.5	11.1
32. Coarse Coal	22.2	21.5	50.4	28.1

The coal above the "holing stone" is not found at the part whence these coals were taken, as before explained. At the N.W. side of the old workings it is three feet thick and has this composition:

	Dawson.
Moisture	1.550
Volatile combustible matter	27.988
Fixed carbon	60.837
Ashes	9.625
	100.000

In these assays we have a most instructive and interesting set of experiments, the most complete of the kind, so far as I know, ever made on any bed of coal of considerable thickness. "All the coals afford a fine vesicular coke and their ashes are light grey and powdery, with the exception of those of the coarse coals which are heavy and shaly. The worst defect of this coal is its containing rather a large quantity of light bulky ashes which causes it to be less esteemed for domestic use than on other grounds it deserves. It is very free from sulphur, burns long, with a great production of heat, and remains alight when the fire becomes low much longer than most other coals." Subsequently to the investigations of Dr. Dawson some abandoned works on the main seam were re-opened and it was reported in 1859 that it was believed "to be generally admitted that the new Mulgrave workings lately opened by Mr. Scott, produce a richer coal than any that has hitherto been obtained from this colliery." The deep seam at these mines, 150 feet vertical below the main seam, contains about 12 feet of good coal, the best of which is superior to that of the main seam. Its best portions contain only about 5.3 to 11 per cent. of ash and afford much illuminating gas and a fine vesicular coke similar to that of the main seam coal. These results were obtained by Dr. Dawson who also tested on a small scale the gas producing qualities of the coal which has long been largely used here and in the United States in the manufacture of gas; the results obtained were :—

	Cubic feet gas per ton.
Coal from upper 9 feet of main seam from Dalhousie pits	3,902
Coal from middle of main seam	5,080
Coal from upper 3 feet of best coal of deep seam	6,668
Coal from lower 3 feet of best coal of deep seam	8,504

Dr. Dawson anticipated that the reputation of the coal as a gas producer might increase as the quality of the coal then being opened was superior to the old, and mentioned that the value of the coal for the purpose in question as well as for family and steam uses depends in part on the good quality of its coke, and in part on its comparative freedom from **sulphur**. It appears that the character of the coal is well maintained. Mr. Hudson, the present **agent at the Albion Mines**, has kindly furnished me with the following statement from the Boston Gas Light Company, U. S., dated Sept. 24th, 1867: "We have not carbonized much of the Pictou Coal received during the present season; the general appearances however indicate that it will **average with former years.** Pictou coal yields 7,180 feet to the **ton of 15 candle gas**; the coke is light and of fair quality. The coal is valuable to such as require large stocks on hand as it is not liable to spontaneous combustion under any ordinary conditions of storage. When the coal is worked to high heat it yields more gas, but to the injury of its coke; signed, W. W. Greenhough." Mr. Hudson writes "I enclose you a copy of letter from the Manager of the Boston Gas Co.—they do not extract the whole of the gas as the coke from our coal, on account of its freedom from pyrites, is of great value; 15 candle gas is burnt in Boston from our coal, in London it is 12, and in Manchester only 10, here the finest gas coal in England is mined, viz: the Wigan cannel of Lancashire. Albion Mine coal is also used for iron making and steam." Mr. Buist, of the **Halifax Gas Works**, found the Pictou coal to give 8,000 cubic feet gas to the ton of 2240 lbs. It will be interesting to have for comparison on the point of gas quality a few lines from the official report of Dr. Letheby on the gas supplied to the city of London in the course of the quarter ending 30th Nov., 1867. It appears that while in the quarter named and the corresponding quarter of the preceding year the average of the gas was about 14 candles, or two above the parliamentary requirement, the chemical quality of the gas as regards the amount of sulphur contained, with one gas excepted, was unsatisfactory, the quantity of the sulphur being generally in excess of the amount allowed viz. 20 grains in the 100 cubic feet of gas. The freedom from sulphur in the Pictou coal is obviously an important matter.

Coal and Oil Coal of the Acadia Mines.—I am indebted to Mr.

Hoyt, Agent of the company working the mines, for much valuable information and have myself examined the oil coal and compared it with corresponding minerals in a paper published several years ago. The McGregor seam of bituminous coal, 12 feet thick, consists of two benches separated by a slaty band which it is difficult to keep from the good coal the character of which is thereby injuriously affected. The two benches also differ in quality as shown in the following analyses obtained by Mr. Hoyt from the former proprietor, Mr. J. D. B. Fraser.

	1st Bench.	2nd Bench.
Volatile matter	22.50	23.30
Fixed carbon	65.70	70.00
Grey ash	11.80	6.70
	100.00	100.00
Specific gravity	1.334	1.301
Theoretical evaporative power, 1 find to be	9.02	9.51

On the 9th of February 1865 one ton of this coal, a mixture from both benches, was tested at the works of the Manhattan Gas Company, New York, with the following results, viz:—

One ton of 2240 lbs. gave 9500 feet of 13.03 candle gas and 41 bushels of coke weighing 1640 lbs. The coke was good, contained rather much ash and made some clinker, but it burned well and kept up a good strong fire. The coal seemed to deserve trial on a larger scale as it was very readily carbonized, yielding a good volume of gas and coke. Analysis of the coal gave:

Volatile matter	32.0
Fixed carbon	59.3
Ash	8.7
	100.0

A subsequent trial made by the same company showed a less favorable result caused wholly by the admixture of the slaty band material. Mr. Hoyt expresses the opinion that if this is carefully thrown aside the McGregor coal would be found superior for gas to that of many other mines now so extensively used throughout New England. The slack coal does not answer for blacksmiths' purposes but a successful experiment has been made on its conversion into coke, samples of which have been approved in various local markets.

The Acadia seam is 20 feet in thickness and is described as "one of the finest seams of coal in the world from the peculiar character of its position and construction. With the exception of the fireclay parting of three inches thickness there is not the least interstratification of foreign matter or impurity of any kind, and the parting is of great importance to the miner because it enables him to "hole" in it, instead of cutting the hard coal, and thus reduce the expense of mining at least 15 cents per ton over any other coal in the country. The Acadia coal has acquired so high a reputation as a superior household coal that it was found necessary to issue certificates with each load sold as a protective measure. As a steam coal it is believed to be the best in America and to equal that of Wales. Vice Admiral Sir James Hope was induced to test a few tons and he was so well pleased with the result that he made a report on it to the Lords Commissioners of the Admiralty. He requested permission to order 1000 tons of the coal and asked the price at which it could be delivered at Bermuda, Jamaica, and Barbadoes. By the report of the Admiral's engineer it would appear that the coal burns quickly, gives a stong heat, makes little or no smoke and is of a hard nature calculated to stand transhipment without deterioration. The coal contains 77 per cent. of carbon, which is only 7 per cent. less than the average of 37 Welch coals, and 5 per cent. less than 18 Newcastle coals. The sample tested by the Admiral was mined only 33 yards from the crop, and might be called crop coal, and was by no means equal in quality with that which could be produced from a further depth. The importance of possessing so superior a steam coal can hardly be overrated. Already applications have been received from prominent shipping merchants for sample cargoes for the West Indian and South American markets which are now supplied direct from Wales. In addition to the British Navy, the steamship companies must inevitably use this coal in place of Welch which they are now compelled to import at a heavy outlay for freight."

I examined a small sample of coal selected by myself from the heap at the mouth of the pit opened by Mr. J. Campbell, at Bear Creek, on the seam said in the Inspector's report of 1866 to be the same as that worked by the Acadia Company just referred to, with the following results:—

Moisture 3.70
Volatile combustible matter........23.94
Fixed carbon ....67.40 } Coke 72.36
Ash............................ 4.96
                                ——————
                                100.00
Theoretical evaporative power 9.26

Oil Coal.—I believe this material was first examined and described by myself in a paper published * in 1860 soon after it had been opened upon by Mr. Fraser. It has been called the Stellar coal from the fact of "stars of fire" dropping from it when it has been held to a flame and removed. The seam in which it is found is called the Stellar seam. As the well known minerals analogous to it in the leading property of furnishing much oil have been distinguished from coals by the special names Torbanite and Albertite this might be designated Stellarite. It occurs with bituminous coal in a seam 5 feet thick of which 1 foot 10 inches are Stellarite, 1 foot 4 inches bituminous coal and 1 foot 10 inches bituminous shale; the composition of the three bands is shewn by my analyses to be as follows:—

	Coal.	Stellarite.	Shale.
Volatile matters.............	33.58	66.56	30.65
Fixed carbon.................	62.09	25.23	10.88
Ash	4.33	8.21	58.47
	100.00	100.00	100.00
Moisture		.23	
Specific gravity.............		1.103	

The oil-coal or stellarite has been examined abroad with quite analogous results, the mineral improves in quality towards the east while the overlying M'Gregor coal deteriorates in that direction. Other analyses have given the following results, the No. 2 is probably the shale:—

	No. 1		No. 2	
	Penny.	Wallace.	Penny.	Wallace.
Moisture	.20	.32	.80	.60
Volatile combustible matter...	67.26	68.38	34.16	38.69
Fixed carbon.................	24.03	22.35	12.30	8.26
Sulphur	.11	.05	.74	.25
Ash..........................	8.40	8.90	52.00	52.20
	100.00	100.00	100.00	100.00
Specific Gravity.............	1.069	1.079	1.612	1.568

* Silliman's Journal and Edinb. Phil. Journal.

Having, on account of my former connection with the British Admiralty Coal Enquiry, been one of those engaged to furnish chemical evidence in the famous first trial in Edinburgh of the question whether the mineral known as "Boghead Coal" found at Torbane Hill, Linlithgowshire, should properly be called a coal, I was naturally much interested on the discovery of the stellar oil-coal and got ultimate analysis made of it and of the "Albert Coal," also subject of a trial on the ground that it had been improperly called coal. These analysis were very kindly made for me through Prof. Anderson of Glasgow who generously met my deficiency in the necessary apparatus which I had not brought out with me. The results were most interesting, especially when compared with those obtained from bituminous and cannel coals. As to the former I selected, from those I had made in the Admiralty Enquiry, analyses of English, Scotch and Welsh bituminous coals, and, as to the latter, analyses of English and Scotch cannels made by other chemists. The following table shews the differences which obtain between these minerals in proximate and ultimate analysis, and in specific gravity, and the ratio existing between the two most important constituent elements:—

Mineral.	Locality.	Specific Gravity	Proximate Analysis.			Ultimate Analysis.					Ratio of Carbon to Hydrogen.	Authority.
			Volatile Matters.	Fixed Carbon.	Ash.	Carbon.	Hydrog'n	Nitrogen.	Sulphur.	Oxygen.		
Welsh Bitum. Coals.	Duffryn	1.326	15.70	81.04	3.26	88.26	4.66	1.45	1.77	0.60	100 : 4.82	H.How
	Newydd	1.310	25.20	71.56	3.24	84.72	5.76	1.56	1.21	3.52	100 : 6.79	"
	Ebbw Vale	1.275	22.50	76.00	1.50	89.79	5.15	2.16	1.02	0.39	100 : 5.73	"
Scotch Bitum. Coals.	Grangemouth	1.290	43.40	53.08	3.52	79.85	5.23	1.35	1.42	8.58	100 : 6.61	"
	Fordel	1.025	47.97	48.03	4.00	79.58	5.50	1.13	1.46	8.33	100 : 6.93	"
English Bitum. Coals.	Broomhill	1.025	40.80	56.13	3.07	81.70	6.17	1.84	2.85	4.37	100 : 7.55	"
	Lydney	1.283	42.20	47.80	10.00	73.52	5.69	2.04	2.27	6.48	100 : 7.73	"
Eng. Cannel	Wigan	1.276	39.64	57.66	2.70	80.07	5.53	2.12	1.50	8.08	100 : 6.90	Vaux.
Scotch Cannels.	Lesmahagow	1.251	56.70	37.26	6.03	73.44	7.62		1.14	*	100 : 10.43	Miller.
	Capledrae				25.40	56.70	6.80	1.90	0.35	8.80	100 : 11.99	A.Fyfe
Torbanite	Torbanehill Scotland	1.170	71.17	7.65	21.18	66.00	8.58	0.55	0.70	2.99	100 : 13.00	H.How
Albertite.	Hillsboro, New Bruns.	1.091	54.39	45.44	0.17	87.25	9.62	1.75		†	100 : 11.02	Slessor & How
Stellarite.	N. Glasgow Nova Scotia	1.103	66.53	25.23	8.21	80.96	10.15	0.68		‡	100 : 12.53	"

* Nitrogen and Oxygen 11.76. † Sulphur if any, and Oxygen, 1.21. ‡ N, S, and Oxygen, .68.

In the paper in question I pointed out that the true comparative value of combustible minerals, while partly indicated by the relative amounts of volatile matter and fixed carbon, is only truly shewn when account is taken of the oxygen, which is sometimes large in quantity, as seen above, and is reckoned as volatile matter to the credit of the mineral while its real effect is reduction of value. I showed that when the hydrogen equal to the oxygen present is deducted, taking only those cases where there is an apparent equality in the ratio of carbon to hydrogen, the last three minerals in the table above stand apart from the rest: thus

Cannel coal from Wigan..................100 to 5.65
„   „   „   Lesmahagow............100 „ 8.71*
„   „   „   Capledrae..............100 „ 10.05
Torbanite   „   Scotland................100 „ 12.43
Albertite   „   New Brunswick..........100 „ 10.85
Stellarite   „   Nova Scotia............100 „ 12.43

* Allowing two per cent. for nitrogen.

and that theoretically they should be excellent "oil coals," as is abundantly shewn by experience. For the following amounts of oil yielded by various materials I am indebted in part to Mr. Poole formerly manager of the Fraser Oil-coal Works where the stellarite was used and in part to Mr. Hoyt. I have myself tried none of them for the production of oil.

Crude Oil per ton.

Union Oil Coal of West Virginia affords......... 32 gallons.
Elk River „   „   „   „   „   54   „
Kanawha „   „   „   „   „   88   „
Lesmahagow Cannel, Scotland   „   40   „
Albertite, New Brunswick   „   ....92 to 100   „
Torbanite, Scotland   „   ..116 to 125   „
Stellarite, or "Stellar Coal" „   53   „
    „   „   „   No. 2..50, 60½, 63, 65, 74 „
    „   „   „   No. 1.........123 to 126   „
    „   „   „   Picked samples gave
                                     in Boston..........199   „

Some of these are the amounts yielded by careful experiments on the small scale: when oil was made at the Fraser mines in 1859 the practical result was about 60 gallons crude, and from 30 to 35 gallons fine clarified oil to the ton. A seam of oil coal similar to that just described was worked for two years about three miles to the east of the Albion Mines. A specimen of oil coal having very much the appearance of the stellarite was sent to the Paris

Ratio of carbon to hydrogen after deducting hydrogen equal to oxygen present.

Exhibition, it was found, I believe, on the east side of East River. The oil coal has also been met with on the Montreal and Picton area, on the northern edge of the basin. At some future time, therefore, great results may be expected from the working of this valuable material. The raising of this mineral was stopped on the discovery of the abundant supplies of mineral oil in the United States about 1860. The whole quantity sold from the two mines in operation was about 4000 tons, of the value of about $8.35 per ton delivered at the place of shipment; some of this was sent to oil-works in United States. Of course the stellarite is a most valuable gas material, it has been used by various establishments in these provinces to mix with bituminous coals for adding to the illuminating quality of the gas produced. Torbanite or "Boghead coal" has been, and is still probably, imported for the same purpose.

Coal of the Montreal and Picton Mines.—I examined several samples of the coals raised on the first opening of the seams; the following is an abstract of my report made to the company as respects the qualities of the coals:—

Sample No. 1, from the first bench, gave
  Moisture 4.40
  Volatile combustible matter 24.95
  Fixed carbon .... 61.07 ⎫ Coke 70.65
  Ash .... 9.58 ⎭
  ─────
  100.00

Theoretical evaporative power 8.39

This coal has considerable evaporative and heating power, and would give a moderate amount of gas of good illuminating quality. The appearance of the coal is much in its favour, some that I saw taken from the seam was very clean and bright.

Sample No. 2, from the second bench, gave
  Moisture 5.47
  Volatile combustible matter 19.93
  Fixed carbon .... 68.55 ⎫ Coke 74.60
  Ash .... 6.05 ⎭
  ─────
  100.00

Theoretical evaporative power 9.41
Specific gravity 1.36

This was an extremely bright and clean coal. Its very high evaporative power makes it occupy a very good position among British and American coals for steam purposes.

So far as comparison can at present be made, there seems reason to believe that the opinion held as to the coal tending to improve in quality towards the northern edge of the basin, is correct. In the specimens examined I observed the sulphur to be small in amount and in this important respect like other Pictou coals they compare favourable with Sydney (and most other Cape Breton) coals for gas and steam uses. Taking all the characters of the coal into consideration, I think there is great promise of their proving respectively valuable for gas, steam, blacksmith's and household purposes.

"A trial of the coal from the third bench made at the Pictou Gas Works, gave 62½ per cent more gas than was, according to Dawson, produced from Albion Mines main seam coal at the same works, and a five foot burner gave a light equal to 15 candles." (Company's Prospectus).

Coal of the Pictou Mines, on the east side of East River, taken from the four feet seam, was tested in Portland, where it was called excellent for steam purposes and found to yield nearly as much gas as a mixture of half Westmorland (U. S.) and half Newcastle coal. Tried at the Pictou Gas Works it gave 6650 cubic feet per ton. The oil coal found here proved in Portland to equal the best cannel coal for gas making.

Oil Coal of Antigonishe County.—Mr. J. Campbell reports a five foot seam of curly cannel which will yield at least 40 gallons of crude oil to the ton, and 15 feet of oil shale which will yield at least twenty gallons.

Coal of Inverness County, Cape Breton. I examined specimens taken as the average of several pounds weight from two seams at Chimney Corner where operations have lately been begun on some seams cropping out on the sea shore. The results were these:—

	No. 1.	No. 2.
Moisture	8.19	7.97
Volatile combustible matter	26.39	20.46
Fixed carbon	57.70	61.13
Ash	7.72	40.44
	100.00	100.00
Theoretical evaporative power	7.89	8.39

The amount of sulphur was apparently about that usual in coals of this district, decided, but not excessive, both samples caked, the first swelled a little and the coke, 65.42 in the first, and 71.57 in the second, was coherent. The samples were of course from near the crop and are promising from their very fair heating power: the coal would prove, no doubt, excellent for domestic purposes.

Coal of Richmond County, Cape Breton. The coal now worked by the Richmond Company, at Little River, has been examined by Dr. Dawson with these results:—

Volatile matter	30.25
Fixed carbon	56.40
Ash	13.35
	100.00
Specific gravity	1.38

The coal is described as hard and very little injured by exposure: "burned in a stove it ignites readily, fuses, swells, and cakes, giving a strong flame and a lasting fire. In a smith's forge it works well, its behaviour being similar to that of Pictou coal. Practically it will be found serviceable for domestic use, well adapted for smith's use, and from the large quantity and high illuminating power of its gaseous matter, probably a good gas coal. There should be little waste in its extraction and it will suffer little by being "banked" or kept in the open air. It contains more sulphur than the Pictou coal."

The coal at Carribou Cove in the same district gave Dr. Dawson

Volatile matter	25.2
Fixed carbon	44.7
Ash	30.1
	10.000

and is described as too poor for exportation but possibly useful for local domestic fires.

Of the Sea Coal Bay coal it is stated in the Inspector's report of 1864 (by Mr. J. Campbell.) "The large bed of coal on which the government received an unfavorable report some years ago has been explored by means of a shaft to the depth of 55 feet, and at that depth the lead is found to improve so much that at least six feet of good clean coal may be mined from it," and "there is another important feature in this coal field. The beds are found to improve greatly in the quality of their coal the further they are followed along their strike to the south-westward and also to the dip."

Coal of Cape Breton County, C. B.—The character of the coal of the Sydney Mines has been well known in the market since the commencement of operations by the G. M. A. in 1827. It is esteemed very highly as a domestic coal and is much in demand for marine steam purposes. Examined by Prof. Johnson on the occasion beforementioned, for the U. S. government it gave the results under his name :—

	Johnson, 1812	How, 1861.
Moisture	3.125	31.87
Volatile combustible matter	23.810	
Fixed carbon	67.570	64.59
Ash	5.495	3.54
	100.000	100.00
Specific gravity	1.310	
Weight of cubic foot in merchantable state	47.44	
,, ,, ,, ,, unbroken, calculated from Sp. Gr., about	83.50	
Actual evaporative power	7.99	
Theoretical ,, ,, I find to be	9.29	8.87

"It burned with a large smoky flame, its coke fell into small pieces and wasted between the grate bars. Its clinker was black compact, and in thin sheets highly fusible, spreading over and adhering to grate bars, heating them. The coal ignited **promptly and burned rapidly**; agglutinating and swelling but slightly."

In 1860 the Inspector reported, "It may be interesting to know that a cargo of 450 tons of this coal was shipped by order of the French Government, to Brest. The Director of the Naval Construction at Brest reports the result of the trials made upon it to the Minister of Marine, as follows : The trials of the Nova Scotia coal by La Perdrix shew that, like Newcastle coal, it ignites easily, and produces a long, lively flame, little coloured. It swells a little in the fire and does not clog the bars. It gives but little clinker, and is not very brittle. Its steam power is little inferior to that of Cardiff coal, and equals that of Newcastle. It is a fine coal and completely assimilable to that of Newcastle." The whole number of calls of steamers, including French, English and American men-of-war, for coals that year was 66.

The quantity of gas yielded is less than that from Pictou coal, and the presence of a little iron pyrites renders it less profitable for the manufacture of gas. As a domestic coal it is quoted in Halifax at the present time at $1 a chaldron more than that from the new mines.

The coal of Lingan was first mined in 1854 for certain gas works in the United States where it was much esteemed. According to Dr. Dawson the 9 feet seam gives a fine coking coal, having a very small per centage of ash and yielding 35.16 per cent of volatile combustible matter, and therefore rather superior to the produce of the Albion Mine as a gas coal. In 1861 the chief portion of the coal raised went to New York, smaller quantities, however, were sent to the gas works at Salem, Boston, Portland, Dorchester, Portsmouth and Quebec, at all which places it was much esteemed. It also makes a good domestic coal, but generally brings 50 cents less in the market than that of Sydney.

Coal of Little Glace Bay Company's Mine.—Coal from the Hub and Harbour veins were thus reported on by the Chief Engineer of H. M. S. Duncan in 1867. "In compliance with your directions to try the two samples of Little Glace Bay, C. B., coals with a view to ascertain their steaming capabilities I have the honor to report that they have had a fair trial in the boiler of the small portable engine attached to the lathe-room and also on board the gunboat Charger. I have also tested them in the usual way for carbon with the following results. The Harbour vein gave dark brown smoke considerable in quantity, 6.79 of clinker and 2.12 of ash per cent., the Hub vein a considerable quantity of light brown smoke, 2.47 of clinker, and 1.3 ash per cent. Both coals light up quickly, raise steam fast, burn well and clearly, and generate steam well. The deposit of soot is considerable in both kinds. Tested for carbon the Harbour vein contains 83.5 per cent. and the Hub vein 80.9 per cent. and therefore in this respect they are nearly equal to Welch which is further corroborated by the fact that the daily expenditure of Welch and Glace Bay coal in the lathe-room is as nearly as possible alike, the Harbour vein having slightly the advantage of the Hub vein. Not having the necessary apparatus, I am unable to test these coals for sulphur, but judging by the manner in which they burn and other observations I should say the quantity they contain is very small. Being similar in their nature to North of England coal they are not liable to make much small or dust and would therefore stand the effects of transhipment without much deterioration. I am therefore of opinion that both of these coals are well suited for the sue of H. M. Steam Ships particularly if treated in the same way as ordered by the Admiralty respecting English coals, viz. mixed with Welch in proper proportions : signed, Edw. O. Crichton."

Mr. Hendry reported in 1864 that the coal stood high in the market both for gas and domestic purposes. A specimen of the Hub vein coal gave Mr. H. Poole, while a pupil of mine, these results.

Moisture	5.52
Volatile combustible matter	31.02
Fixed carbon	62.53
Ash	.93
	100.00

Theoretical evaporative power............. 8.59

Mr. Buist has found one of the coals of Glace Bay to yield at the Halifax Gas Works 8500 cubic feet of gas to the ton of 2240 lbs. The coal contains more sulphur than Pictou coal but the gas is superior in illuminating quality, its coke is very good. Of the two seams Mr. Buist says the other is not so well adapted to gas purposes but is very superior for steam purposes.

Coal of Gowrie Mines.—The following report is from Mr. W. T. Rickard, F. C. S. "Having carefully analyzed samples of coal from Gowrie Mines, Cow Bay, C. B., I beg to observe the general properties of the coal are of a highly satisfactory character, and for steam and domestic purposes may be considered equal and in some respects superior to Newcastle coal. It burns with a strong hard cinder, giving off very little smoke, and leaving an ash which is not liable to form clinker owing to the absence of lime. While the proportion of sulphur is not excessive in any of the samples, in that from the second band it is less than the average obtained from 36 samples of Welsh coal. Its evaporative power, though inferior to Welsh, is higher than that of Newcastle coal, while its high specific gravity renders its capacity for storage not the least of the advantages it offers for Marine Steam purposes. The results obtained on analysis, in 1863, were given as follows:—

	Sample marked First and second Band.	Sample marked Second Band
Moisture	1.80	1.46
Hydrocarbonaceous matter	27.08	37.27
Sulphur	3.42	1.27
Coke	67.70	60.00
	100.00	100.00
Ash	7.25	4.15
Theoretical evaporative power	8.53	
Specific gravity		1.33

The coal of the Block House Company is much in request for gas in the cities of New York and Boston. Mr. Buist informs me that he has found the coal to give at the Halifax Gas Works 8,500 cubic feet of gas to the ton of 2240 lb. This coal contains more sulphur than that of Pictou, but the illuminating power is a little better; the coke obtained is of fair quality.

Table shewing the Proximate Analysis of some of the Coal and Allied Minerals mined in Nova Scotia:

Locality.	Minerals.	Specific Gravity.	Cubic feet of gas per ton.	Gallons of oil per ton.	Moisture.	Volatile Combust. Matter.	Fixed Carbon.	Ash.	Prac. Evaporative Power.	Theoretical Evaporative Power.	Authority.
Cumberland Co., N. S. Joggins.	Coal.				2.50	36.30	56.00	5.20			J W Dawson.
Springhill, S.	,,				1.80	28.40	56.60	13.20			,,
,, ,,	,,				2.92	22.46	60.95	13.67		8.37	H. How.
,, North. *Pictou Co, NS*	,,				37.00 a	59.17	3.82				,,
Albion Mines.	,,	1.32			2.56	27.06 b	56.98	13.38	8.41	7.82 (H)	Johnston.
,, ,, ,,	,,	1.33			0.78	25.79	60.73	12.51	8.49	8.34 (H)	,,
,, ,, ,,	,,		7180								Greenhough.
,, ,, ,,	,,		8000								Buist.
Acadia ,,	,,	1.334	9560		22.50	65.70	11.80		9.02 (H)		,,
,, ,, ,,	,,	1.301			23.30	70.00	6.70		9.51 (H)		,,
Bear Creek.	,,				3.70	23.94	67.40	4.96		9.26	H. How.
Acadia Mine.	Oil coal or Stellarite, No 1.	1.103			0.23	66.56	25.23	8.21			
,, ,,	,,	1.069		123	0.20	67.26 c	24.03	8.40			Penny.
,, ,,	,,	1.079		126	0.32	68.38 d	22.35	8.90			Wallace.
,, ,,	No 2.	1.612		60	0.80	34.16 e	12.30	52.00			Penny.
,, ,,	,,	1.568		63	0.60	38.69 f	8.26	52.20			Wallace.
,, ,,	Oil Shale.				30.65		10.88	58.47			H. How.

COAL AND ALLIED MINERALS.

(Continued.)

Locality.	Mineral.	Specific gravity.	Cubic feet of gas per ton.	Gallons of oil per ton.	Per cent. Moisture	Volatile Combust. Matter.	Fixed Carbon.	Ash.	Prac. Evaporative Power.	Theoretical Evaporative Power.	Authority.
Pictou Co,NS											
Acadia Mine.	Coal in Stellar sm		....	....		33.58	62.09	4.33	....	....	H. How.
Mont & Pictou Mine.	Coal.		......	....	4.40	24.95	61.07	9.58	....	8.39	,,
do.	,,	1.36	......	....	5.47	19.93	68.55	6.05	....	9.41	,,
Pictou Mine.	,,	......	6650	....	....		....	...	....	....	Pic.Gas Wks
Inverness Co., C. B.											
Chimney Cor.	,,	......	......	....	8.19	26.39	57.70	7.72	....	7.89	H. How.
,, ,,	,,	......	......	....	7.97	20.46	61.13	10.44	....	8.39	,,
Richmond Co. C. B.											
Richmond.	,,	1.38	......	....		30.25	56.40	13.35	....	....	J W Dawson.
Cape Breton Co., C. B.											
Sydney Mine.	,,	1.34	......	....	3.12	23.81	67.57	5.49	7.99	9.29 (H)	Johnson.
,, ,,	,,	......	......	....		31.87	64.59	3.54	....	8.87	H. How.
Lingan ,,	,,	......	......	....		35.16	....	....	....	....	J W Dawson.
Lit. Glace Bay Hub Vein.	,,	......	......	....	5.52	31.02	62.53	0.93	....	8.59 (H)	H. Poole.
,, ,,	,,							1.30			Crichton
do Harb'r. do.	,,	?	8500	....	....		....	2.12	....	....	and Buist.
Gowrie, 1st & 2nd Band.	,,	......	......	....	1 80	30.50 *g*	60.45	7.25	....	8.53	Rickard
do. 2nd Band.	,,	1.33	78500	....	1.46	38.54 *h*	55.58	4.15	....	....	and Buist.

a Sulphur 0.316. *b* Sulphur 0.769. *c* Sulphur 0.11. *d* Sulphur 0.05. *e* Sulphur 0.74. *f* Sulphur 0.25. *g* Sulphur 3.42. *h* Sulphur 1.27.

Oil Shales.—Up to the present time no attempt has been made to manufacture oil from shale, except in the case of that accompanying the Fraser oil coal, both of which were worked together I believe in the distillation of oil as already mentioned. Large deposits of shale are met with in the East River district of Pictou county affording a product of oil larger than is found remunerative in Scotland. Last year I observed an advertisement in a Glasgow paper referring to a shale yielding about 30 gallons of oil to the ton for which offers were invited, and some shale I examined from East River gave about 35 gallons crude oil to the ton. The non-productive coal measures of Hants county afford large quantities of shale, which have led to expectations of finding coal, but the amount of oil they yield has not been ascertained. The deposits of shale in Antigonish county may be of the same age as these; these beds are very favourably spoken of by Mr. Campbell, from whose report I make a few extracts. "The fact that the centre of the Antigonish basin is occupied by highly bituminous limestone overlying the oil-coal and oil-shale beds may possibly indicate that the whole group is upper devonian or lower carboniferous rocks which are not known in this country to contain coal beds of any value." On this point I may mention that in a depth of about 180 feet in the neighbourhood of Windsor only one small seam of coal, some six inches thick, was found in 1864, and that in a shaft sunk at Hantsport in similar rocks to a considerable depth no coal was obtained. Mr. Campbell goes on to say : "The bituminous beds appear to be divided into two groups, the lower of which appears to be about 70 or 80 feet in thickness, 20 feet of which may be regarded as good oil shale including five feet of curly cannel rich in oil. The upper band, which lies in immediate contact with the limestone, cannot be much short of 150 feet in vertical thickness of strata containing a large percentage of oil. Of this great bed of oil-batt about 30 feet will in all probability yield from 20 to 25 gallons to the ton. The five feet seam of curly cannel will yield at least 40 gallons crude oil to the ton, and the fifteen feet of the best section of the oil-batt will yield at least 20 gallons to the ton, and taking this as worth 25 cents per gallon at the shipping port, there are in all $370,533,325 worth of oil which can be obtained from 20 feet in thickness of strata underlying 2000 acres of land—out of 18000—comprising a basin underlaid by at least 50 feet in thickness of beds rich in oil."

Bitumen.—Liquid Bitumen, petroleum, or mineral oil, is reported to have been found in more localities than one but I have no precise information on this subject. A most interesting discovery of solid bitumen was made by Mr. Barnes in the neighbourhood of Grand Anse, Inverness county; the amount met with was small. The remarkable resemblance between this bitumen and the albertite of New Brunswick was pointed out in a paper by myself (Contributions to the Mineralogy of Nova Scotia, II. L. E. D., Phil. Mag. March 1866). Bituminous matters are often found diffused through limestones, sandstones, and occasionally gypsum, but separate masses such as observed by Mr. Barnes have not before been noticed.

Peat.—Extensive peat bogs are found in the province; the largest perhaps are the savannahs in different parts of Shelburne county, and the carribou bog near Aylesford, Kings county. They are especially frequent in the rocky country of the Atlantic coast, as near the gold districts of Tangier and Sherbrooke. Mr. Campbell mentions the existence of considerable tracts of peat on the table lands in the Cape North district of Inverness and Victoria counties, Cape Breton. No examination has been made of the depth and quality of the peat in these deposits but the subject is well worth attention as peat is now extensively used as a fuel in generating steam, and in smelting and working iron, especially in the case of iron-sand, which, as hereafter mentioned, is met with on the Atlantic coast of the province. (See geology of Canada 1866, p. 284, et seq).

CHAPTER III.

GOLD.

A GLANCE at the Introduction will show that a good deal has been written on the auriferous rocks of the province since the commencement of gold mining in 1861. The results of all but the earliest operations are registered with precision and fulness, very much appreciated abroad, in the official reports annually given at first by the Chief Gold Commissioners and latterly by the Chief Commissioners of Mines. A summary only of the numerous and very interesting details on record is necessary on the present occasion; with this will be incorporated such additions as seem useful.

Gold was found in Nova Scotia about 100 years ago. Not long after my arrival in the province, probably in the summer of 1855, the late Canon Gray, D.D., Rector of Trinity Church, St. John, N. B., who died in 1868, aged 70, told me that as a boy he had taken gold out of rocks on his father's property near Halifax, and had it melted by a jeweller in that town. On my speaking of this lately to his son, B. G. Gray, Esq., this gentleman said the locality referred to was probably near Bedford, and that both his grandfather and great grandfather had collected specimens of gold in this province; that the latter especially had many specimens of rich auriferous quartz from Lawrencetown and other now well known localities, and that old family documents show that particular importance was attached to certain parts of the estate, presumably from the known existence of gold.* In the historical account of the discoveries of gold in Mr. Heatherington's guide to the Gold Fields, it appears that the metal was known to exist in Sherbrooke, Isaac's Harbour, and the Ovens, more than 30 years ago, that Mr. R. G. Fraser and Mr. J. Campbell found alluvial gold in 1857, that Capt. L'Estrange discovered gold in quartz in 1858, and that Mr. J. G. Pulsiver, in

---

* I have been informed since the text was written that gold was washed from the Avon at Windsor, Hants, sixty years ago, each man making eighteen pence a day.

1860, made that discovery which led to the opening up of Tangier, and the eventual proclamation of all the other districts.

Gold is found in Nova Scotia in the metamorphic district forming the southern half of the province, in the altered rocks of different parts of the northern and eastern portions of Nova Scotia and Cape Breton and in the detrital deposits derived from these rocks, respectively, met with in lower carboniferous formations; it is said to occur also in the trap of the new red sandstone, and is often found in superficial alluvial deposits.

Almost all the gold has been mined in districts lying scattered through the band of metamorphic rocks considered to be of lower silurian age extending through the southern half of Nova Scotia proper from Cape Sable at the west to Cape Canseau at the east. This zone varies in width from 6 or 8 miles at its eastern extremity to 40 or 50 at its widest parts, its edges being only roughly parrallel. It is estimated to contain about 7000 square miles. The rocks consist of quartzite and slates, inclined at a high angle, with a general northeast and southwest course; they are invaded in places by granitic eruptive rocks and in the vicinity of these are dislocated and highly metamorphosed. They seldom rise to any great height, the highest land on the Alantic Coast is Mount Aspatogoen, ascertained by Mr. Poole's measurement to have an altitude of 450 feet. Mr. Campbell reports that from the entrance to Halifax Harbour to the Renfrew Gold Field, a distance of about 30 miles, six nearly parallel anticlinal folds or lines of elevation exist, which do not lie at equal distances apart because the strata are folded up to sharper lines of inclination in some of them. The east and west anticlinal lines are intersected by north and south lines of upheaval and it is of the utmost importance to find the east points of occurrence of these as it is chiefly at such localities that the gold bearing rocks are brought to the surface. The band nearest the sea contains, in the east, the Ovens and Tangier mines; the next has Wine Harbour and Lawrencetown; the third, Sherbrooke, Old Tangier and Birch Cove; the fourth, Waverly and Isaac's Harbour; the fifth, Oldham and County Harbour; the last, Renfrew gold field. It is stated by Mr. Hamilton, in a paper read while he was Chief Gold Commissioner, that the bands do not in every instance continue from one gold bearing tract to another, and he is of opinion that the quartzite bands are much more numerous than they have been represented. In the western part of the province gold has not been mined west of Lunenburg County, but it has been met with at Middle Jebogue and the Cream Pot, in Yarmouth county at the extreme west of this southern metamorphic district. In the eastern part of the band it has been found at the farthest extremity on the shore of Chedabucto Bay. New discoveries are continually being made throughout the region and it is thought probable that the metal has been found wherever it has been sought in the proper rocks from one end to the other of the southern part of the province.

The gold exists to some extent in the quartzite and slates but generally occurs in leads consisting chiefly of quartz running through the quartzite and less frequently through the slates. The leads contain more or less mispickel, galena, blende, iron pyrites, copper pyrites, and peroxide of iron; argentiferous copper sulphide, molybdenum, and, as at Oldham, native copper are occasionally observed. The gold is frequently alone, in a state of minute subdivision and invisible, also in masses of characteristic appearance varying in size from scarcely visible points up to pieces, of irregular form, of a few ounces in weight: crystals are met with but rarely, I believe. The most abundant sulphuret is probably mispickel, this sometimes occurs in very large masses, as at Montague where it frequently holds gold in visible particles; the same association is seen at Uniacke. In some leads other sulphurets are considered good indicators of gold, as blende in one of the leads at Waverly; at other leads in the same locality it is not found to be so. Native sulphur in crystals has been found at Uniacke as mentioned to me by Prof. Hind. Of silicious minerals in the quartz, chlorite is found at Uniacke and, as Dr. Lawson informs me, soda felspar at Waverly. The slates are generally argillites, at Tangier and to the east they are magnesian, at Wine Harbour the gold occurs in talcose slate which is sometimes beautifully plated with the metal.

As a general rule those leads only are found to be productive which are conformable with the containing rocks and are therefore held to be beds as distinguished from veins. Notable exceptions occur at the Ovens, and at Oldham, where very rich cross veins are found.

The words vein, lead, and lode are applied indiscriminately to beds and true veins. The productive leads vary in thickness from a small fraction of an inch up to several feet, they generally, if not always occur in groups. The arrangement of the groups is so regular in some places that if the outcrop of the highest or most recent is known all the others may be found in proper sequence beneath it. As many as 30 veins ranging from an inch to 15 inches in thickness have been found in a distance of 160 feet, and at Lawrencetown 5 distinct groups, three of them consisting respectively 10, 13, and 7 auriferous leads, have been recognized; one group appears to be contained within a space of 45 feet, another within 70 feet. These groups are separated by bands of barren rock and are succeeded by barren leads. The amount of quartz in leads is enormous, for example, the 30 veins above mentioned gave an aggregate thickness of 15 feet, in another case an aggregate thickness of 25 feet has been observed in 250 feet distance, and the greater part of the veins were proved to be auriferous.

The leads sometimes consist of beautifuly white quartz with coarse gold almost free from sulphides, sometimes of much slate with the quartz, when they are often very productive. The upper wall is generally quartzite, the lower slate, both walls occasionally slate, but cases of quartzite forming both walls do not appear to have been observed.* The contrast remarked in Australia between the continuance of hard and soft veins has been noticed here in some cases, viz: that leads in which the quartz is enclosed in a soft friable rock are erratic and do not seem to extend far longitudinally, while those in which the vein stone is very hard are very regular and of long continuation and contain the metal pretty equally distributed. As a general rule it is found that the variation in richness in different points in the length of veins is very great, so that lodes rich on the whole are very poor in some parts, thus for example, when on a visit to Waverly last year Mr. Burkner told me that the Tudor lead was richest in the centre, that at shaft No. 11 the quartz gave from 6 to 8 ounces of gold to the ton, at No. 13 the slate was so rich as to seem all gold in the hanging wall, while at Nos. 17 and 18 there was as little as from 8 down to 3 dwt. to the ton. No mention was made of the depth at which these results were obtained, and the depth is a very important consideration as it appears that here as elsewhere the metal runs

* As these pages are passing through the press **Dr.** Lawson mentions to me that there is such a case at the Lawrence Co.'s mine, at Uniacke.

through the leads in streaks, bands, or zones, which are sometimes horizontal but often inclined at an angle varying from that of the dip of the lead. These rich belts are separated by intervals of barren lead of varying thickness. On the same lead, as far as experience goes, the depth of the rich zone and the distance of the zones apart is tolerably uniform, on a neighbouring lead it is not necessarily on the same horizon but may be much below or above. (See Mining Gazette, April and May 1868). Of these streaks or zones Rittel says, in his Resources of California, that the rich streak has a dip in the lode which may be found by taking some of the vein stone and examining the wall rock carefully. In most veins it will be found that the wall has little furrows, as though the lode had been pushed upwards; these indicate the direction of the dip of the rich streaks. Sometimes the gold occurs in nests and pockets apparently distributed without rule, at other times it is more uniformly disseminated. In cross sections of veins there is also great diversity: in one case the metal is nearly all on one side of the vein; in another lode similarly circumstanced it is chiefly on the other side; in a third—these cases are more rare—it forms a plane or leaf in the middle of the lode. Again, it is sometimes found mostly in the slate casing and not in the quartz itself; the ordinary rule appears to be that the metal is scattered through both. (Hamilton on Aurif. Deposits; Trans. N. S. Inst.). The following observations by Mr. J. A. Phillips, in his recent magnificent work on Gold and Silver, show that the foregoing characters belong to auriferous veins generally; they are offered as interesting on the subject of the variable richness in planes and on that of the results of deep mining in Australia and California. "It is generally observed that the widest auriferous veins are not usually the richest and that some of the laminæ running parallel with the enclosing walls are uniformly more productive than others. It therefore not unfrequently happens that a portion of the rock sufficiently rich to be worked with advantage is separated from another comparatively barren band by a distinct heading or false wall. As a general rule these veins are most productive which contain a good deal of sulphides disseminated, although near the surface these have, in nearly every instance, become decomposed, the enclosed gold has become liberated, and the quartz stained of a reddish or brownish colour. When gold occurs in a hard white quartz free of sulphides it is mostly in visible flakes and granules, but such veins, though affording fine specimens, are not often regularly and remuneratively productive. Some of the most steadily remunerative veins are only of moderate size and seldom exhibit visible gold, and this is particularly noticeable in those which, like the Norambagua lead in Grass valley, California, are divided by numerous thin seams of slate into bands of various thickness. It was formerly belived that veins of auriferous quartz become gradually less productive at increasing depth but more extensive experience would tend to show that it is in reality not the case. Gold mines which have for many years been continuously worked in various parts of the world, have fluctuated considerably in richness at different depths but it has not been found that these variations in any way correspond with a gradual impoverishment in the deeper levels. In a communication addressed to Sir R. Murchison, who inclines to the opinion that gold-bearing veins generally diminish in value in depth, Mr. Sylwin says: "There is undoubtedly good evidence that those upper portions of the quartz veins which by denudation now form the gold drifts were often far richer than any we now find at the surface, but we should not forget that in all probability many hundreds of vertical feet of quartz have been thus naturally broken up, crushed, and washed, and the fact of the veins so abraded being still frequently very rich in their present surface, goes far, I think, to prove that the diminution of yield in depth, even though admitted to be true, on a large scale, is still so slow as not to be appreciable within any depth to which ordinary mining operations are carried. He concludes by expressing an opinion that the extraction of gold from quartz reefs, if properly conducted, may be regarded as an occupation which will prove as permanently profitable in Victoria, as tin and copper mining have been in Great Britain. (The greatest depth to which any reef had been worked in 1861 was about 450 feet, the yield was more than 5 ounces of gold per ton of quartz.)

"In California the early quartz miners were also fully impressed with the idea that the outcrops of the leads were more productive than the deeper portions of the same veins, and as soon as the quartz ceased to pay they usually suspended operations without exploring to any considerable depth. Within the last few years however their opinions in this respect have materially altered since the working of the deeper mines would lead to the conclusion that there is no evidence of the progressive falling off in the richness of the veins in the deeper workings. The North Star and Allison Ranch veins in Grass Valley among many others that might be selected, serve to illustrate the fact that the Californian mines do not become sensibly poorer in depth. The North star is now worked in its dip to a depth of 750 feet, and gives quartz yielding on an average £7 per ton of 2000 lbs., whereas in the upper levels the value did not exceed £4 per ton. The Albion Ranch mine at a depth of 500 feet, yielded during the first three months of 1866 a net profit of above £20,000. Hayward's mine, in Amador county is another still more striking instance; this ledge is worked in its dip to a depth of above 1250 feet and yields quartz of much greater value than that obtained from the same vein at higher levels. Among the reasons for the former prevalence of the impression in question is the fact that gold is almost universally associated with pyrites and other sulphides, and these, becoming oxidised at shallow depths, liberate the enclosed gold, which is thus readily extracted by amalgamation although the deeper and consequently less decomposed portions of the vein, which may in reality have been equally rich, gave less satisfactory results to the early miners. With the improved methods of treatment, however, which have come into general use the difficulty has to a great extent disappeared, and as all the auriferous sulphides are now being carefully collected for subsequent treatment, the average production of a vein has generally been found to be sustained at all depths to which the miner has hitherto penetrated." At the famous Eureka mine of Grass Valley the ore contains only $12 a ton of 2000 lb. at the surface and this only for a small extent of the vein. At 100 feet depth, it contains $25, and at 200 feet depth, $42 per ton. Now the works are at 300 depth, and here a great part of the ore pays $70 per ton: at the same time the vein matter has increased from two feet in thickness at the surface to five feet, (Mining Gazette, Feb. 1868.)

The altered rocks in the inner districts of the province generally considered to be upper silurian or devonian have as yet not afforded much gold but there has been some little mining attempted. These rocks form several ridges. One runs along the south side of the Annapolis Valley from Digby to near Windsor. Another forms the Cobequids and with an alteration in its direction proceeds eastward to the Strait of Canseau throwing off spurs north-eastward to the Gulf of St. Lawrence, and south westward on both sides of the Stewiacke river. In the Island of Cape Breton, nearly the whole of Victoria County, a large portion of Inverness, and several detached eminences in Cape Breton and Richmond counties belong apparently to the same formation. The several ridges have been but little explored for gold, nor is it probable they will be, to any great extent, for sometime to come. These hills are, for the most part, in the interior of the country, their rocks are rarely exposed, being covered with a pretty deep soil bearing a heavy growth of timber. Gold has been found in quartz at Wagamatcook, Victoria county, and mined to some small extent, but little is really known of the economic value of this proclaimed gold district on account of its remote and inaccessible position. In the same county quartz veins yielding gold associated with iron and copper pyrites and argentiferous galena occur in slates and schists near Baddeck, another vein near the junction of a bed of quartzite and coarse hornblendic rock at the same place affords both gold and silver. I examined some specimens from these spots for Mr. Cameron, who afterwards had assays made in Boston the results of which, as reported by Prof. Hind, were that the galena from the slates and schists gave a very remarkable percentage of gold, and the ore from the quartzite gave upwards of 18 oz. of gold to the ton of 2000 lb, and more than 97 oz. of silver. Gold has been found at Cape North, Inverness county, near the head waters of the Musquodoboit river, Halifax county, and of the Stewiacke river, and, it is believed, at Five Islands, Colchester county. Mr. Poole found many quartz veins and other favorable indications in Digby county and gold has been reported from near the town of Digby so that there is reasonable expectation that the metal will be found throughout this metamorphic district.

In connection with the foregoing discoveries it is interesting to observe that in Australia " gold is now found to occur not only in quartz veins and the alluvial deposits derived from these and the surrounding rocks but also in the claystone itself and, contrary to expectation, flat bands of auriferous quartz have been discovered in dykes of diorite which intersect the upper silurian and lower devonian rocks. Quartz of extraordinary richness has been obtained from these bands and the new experience of the miner is leading him to look for gold in places heretofore entirely neglected. It is probable some time may be lost, and that his labours may not be always well directed or successful, but it is commendable that he should not be deterred from exploration by warnings and remonstrances, founded on surmises often baseless. If he had closely followed the older precepts we should, at this moment, have been dependent for our yield of gold on the shallower alluviums and the surface only of the veins of quartz." (R. B. Smyth, Intercolonial Exhibition, 1866, p. 5, quoted in Phillip's Gold and Silver, p. 108.) In Canada the most promising auriferous rocks are upper silurian. In the detritus of lower silurian rocks found forming a conglomerate about 4 miles from Gay's river, Colchester county, in rocks of lower carboniferous age, gold occurs in considerable quantity. The lower part of the beds of conglomerate or grit at their junction with the slates on which they lie unconformably is richly auriferous, the gold occurring chiefly in the form of flattened scales, sometimes a quarter of an inch in diameter, disseminated through the rock. The discoveries made here were chiefly on improved lands and from the high prices asked for permission to work, little has been done in mining, but the official report for 1866 says that it is beyond all question that the conglomerate bed contains a large proportion of gold and it was expected that operations would soon prove its economic value on mining and treating. I have lately heard that very good results are being obtained. A conglomerate derived from the newer metamorphic rocks of Victoria county is reported on by Prof. Hind in connection with the veins before mentioned found by Mr. Cameron. The conglomerate forms a bed about two feet thick underlying limestone, it contains a little copper with remains of plants and also unworn pebbles and masses of crystalline limestone. Dr. Hayes found in these samples: I., 5 dwt. of gold; II., 19 dwt. 14 grs. of gold; III., 16 dwt. 8 grs. of gold to the ton; in the latter case 6 dwt. 12 grs. of silver also. These were the results of assays of selected specimens in which galena and copper ore were visible. It is manifest, the report says, that so easily accessible and rich a conglomerate is very valuable, while the important question remains to be answered whether the gold found on assay is all in a state in which it can be separated by amalgamation and whether the band is equally rich in other quarters. Mr. Poole found a conglomerate, resting unconformably on slates, near Avour's Head, Digby county, which contained gold and native copper.

As regards gold occurring in trap rocks, auriferous quartz has been discovered and to some slight extent mined in the trappean headlands of Partridge Island and Cape d'or in Cumberland county, as stated by Mr. Hamilton. The ridge on the south shore of the Bay of Fundy from Blomidon to Briar Island being of the same nature will doubtless be found to afford it also.

To Mr. J. Campbell appears to be unquestionably due the discovery of the auriferous character of the superficial alluvial deposits of the province. Before 1857 he had panned gold from several places along the sea shore and in that year, aided by Mr. R. G. Fraser, he obtained a very good show of gold from the sands of Fort Lawrence, Halifax Harbor; in the same year, and in 1859, Mr. Campbell and Prof. Silliman obtained gold from the sand of Sable Island. The former reports that in nearly all the deposits of glacial drift or boulder clay on the south coast more or less gold is found, but its economical value is much lessened by its diffusion through tenacious clay, too expensive to work by ordinary means. It is only when the glacial drift has been re-arranged that it will be worth working. The gold of the rocks on this coast is now for the most part probably in the submarine banks as is perhaps sufficiently proved by the sands of Sable Island being richly auriferous.

Shore washings at the Ovens, in Lunenburg county, are reported to have yielded during the autumn of 1861, 2000 ounces, in 1862, 311 ounces, since then they have been abandoned, but Mr. Heatherington states that from his own recent experiments he is able to assert that the sands are still worth testing afresh.

The Chief Gold Commissioner reported in 1863 that both in Wine Harbour and Sherbrooke, but especially in the latter, a large proportion of the material returned as "quartz crushed" in this and the preceding year really consisted of alluvium and *débris* from the pits less highly auriferous than average quartz but very profitable from the facility with which it can be procured and crushed. This year I was informed by Mr. Lordly that the washings at Sherbrooke were becoming very rich and one company had probably taken out 400 ounces of gold. Tangier produced in 1865, 117 ounces from alluvial washings; this and various other districts have returned smaller quantities. Last autumn a deposit of sand was discovered in Mr. Sutherland's claim at Gold River, Lunenburg county, in which I found 14 dwt. 10 grains of gold to the ton of 2000 lb. of dry sand.

In the newer metamorphic districts gold has been found in the alluvium brought down by many streams. In Wagamatcook, which is a proclaimed gold district, most of the gold obtained has been washed from the alluvium on the lower flanks of the hills skirting the Middle river. It is very coarse and nuggetty and is indicative of rich lodes in the high lands. The same rock formation is seen on many places on the shore of the Bras d'or. Gold has been found in the sands of nearly all, if not all, the streams of Inverness and Victoria counties which take their rise in these metamorphic hills.

As full separate returns have not been made for quartz and alluvial gold at Wine Harbour and Sherbrooke it is impossible to ascertain the total amount of the latter obtained, but, taking the foregoing estimate of that got from the Ovens in 1861 and the quantity given in the official returns as "from alluvial mines," it appears that in round numbers 2627 ounces of gold have been obtained from washings up to the date of last returns in Sept. 1867.

It is not possible now to arrive at the actual amount of gold taken out since 1860 because it is well known that no accurate returns were made at first. It was not till 1862 that the Department of Mines was established and although it is binding on both miners and mill owners to make returns under oath of the amount of quartz crushed and of gold obtained the earlier statements are not perfect because it took some time to get the official system of recording in working order. As for the whole it is so well known that a large amount of gold is stolen that it is considered a tenth might fairly be added on this account to the gross amount of gold as given in the official returns, without doing this, and estimating with Mr. Heatherington 6000 oz. as the amount raised in the first two years we have in these and the succeeding years, ending 31st December; the following as the

GROSS YIELD OF GOLD IN NOVA SCOTIA.

Year.	Ounces.	Year.	Ounces.
1860–61	6,000	1865	25,454
1862	7,275	1866	25,204
1863	14,001	1867	27,294
1864	20,023		

Total..................125,251

The gold is valued in the official returns at $18.50 per ounce, hence the value of the total product is $2,317,143.

The various districts which have produced the gold have given very different quantities as shewn in the details to be found in the official returns. The Chief Commissioner of Mines and the Inspector of Mines give a most interesting account of the actual condition of each district in their report for 1867 from which and subsequent returns and other sources I condense the following statement.

In gold mining the success may be considered good both in the increase of gold obtained and the average rate per ton of quartz crushed, whilst the average remuneration for each man, counting 313 days in the year and the gold at $18.50 per oz., is two dollars and forty four cents per day, a result, it is believed, without a parallel in any country. The progress in the yield of gold has been steady and we may expect a large increase in the working of the poorer mines; leads are now made to pay which at first could not have been worked without loss and leads now deemed worthless will no doubt, owing to the increased experience in mining and treating the ores, be found remunerative.

Stormont.—The mining in this district has for some years been confined to Isaac's Harbour, but this year prospecting to a considerable extent has been carried on at Seal Harbour with most promising results. In the County Harbour portions of the district there has been some fairly successful prospecting. This district as a whole has always been a paying one yet from its inaccessability by land carriage it has not hitherto been much known.

Wine Harbour district still shews a falling off, both in the quantity of gold produced and paying qualities per man. This may partly be accounted for by the fact that an extensive tunnel is being cut across the metals for drainage and prospecting purposes preparatory to extensive mining. There have been large purchases made and it may be again, as it was in 1863, the best paying district in the province. New lodes were reported to be discovered in January last. Some of these have lately proved very rich.

Sherbrooke.—Though second in quantity of gold produced, this district, as it has been since **1864**, is the first in profit. The average per man for 1866 had a cash value of $1617.45, and for 1867, $1642.30. Much larger returns are expected this year; extensive preparations are being made for mining.

Tangier taken as a whole has not proved a success, yet the Chief Commissioner cannot believe that in a district where such splendid specimens, so much rich ore, and so many leads are met with, good paying mines will not be found. Old Tangier, or Mooseland, has advanced rapidly and is proving remarkably rich both in ore and specimens. The result of last January's crushing was 64 oz.

Montague is one of those districts that have fallen away since the publication of the last annual report: it is believed the depression will not last; by latest accounts preparations were being made for renewed operations.

Waverley. Though this has lost its place as first in regard to quantity of gold obtained it still continues to occupy a prominent and attractive position among the gold-producing districts. There is no place in the province where, so far as the Chief Commissioner can learn, mining is so economically carried on and crushing so cheaply done as in this district. The great depth of the soil here is believed to be the chief cause of the falling off in returns. Although the yield has been large the width of ground mined has been very small, mining having been confined to one lead. From the narrow strip of ground worked there have been got, up to Sep. 1867, 56758 tons of quartz which have given 36101 oz. of gold. "Can it be supposed that the district has run out when as is well known, from the cause above stated, the district has hardly been prospected at all?"

Oldham has never done so well as in the past year and it may be fairly inferred that the periodical depression to which all districts are so liable has in this instance passed and it will be surprising if in a short time Oldham does not take its place as a leading district.

D

Renfrew, from being third in rank last year, and about fifth in years previous, has placed itself first this year, having produced 900 oz. more gold than any other district, and is second only in point of profit, each man having earned $895.30 for the year. This result may well inspire the miners in depressed districts with confidence. From being one of the poorest **Renfrew** has become in a short time one of the most productive districts. Here the Ophir Company has had great success. The trustees reported: " It is almost unparalleled in the history of mining operations that a mine has been opened to so large an extent, buildings erected, machinery procured, and in fact the whole mining plant paid for, and a very handsome surplus earned above all these preliminary and necessary expenditures, out of the profit of the mine from the start, and without a call of a dollar from the Shareholders in the short space of eight months."

Lawrencetown this year does not figure as a gold producing district, it being classed in the tables among the " Unproclaimed and other Districts;" this is not because there has been nothing done but because there was only a small amount of gold produced. The discoveries of gold bearing leads have been considerable and operations have been commenced on a large scale and there is no doubt from present appearances that this will in future be a leading district. Recent discoveries have induced the commission to enlarge this district to nine times its original size.

Uniacke is a new district in which some prospecting licenses were taken out in 1865. The ground already applied for and under prospecting license and lease is large and the surface over which workable leads have been found equals any in the province. Since the report was issued very large returns have been made. The Mining Gazette for February states that 13 tons of quartz from the Westlake Co.'s mine gave 234 oz. 6 dwt. of gold as the result of 21 days labour from a vein containing from 6 to 8 feet of workable ore opened and being drifted in at a depth of 16 feet. Twenty-five tons of quartz from the same shaft crushed three days before Christmas gave 67 oz. 11 dwt. of gold. From the Hall and M'Alister mines six tons recently crushed gave 22 oz. 7 dwts. retorted gold. The Doull and Burkner properties are also yielding good ore, and prospectors claim to have made discoveries in the western part of the district.

Ovens. This district has not improved and yet the quartz is very rich. The Mining Gazette of February reports that the editor is confident from the results of a visit and tests personally conducted that both the quartz and placer mines of the **Ovens** offer lucrative fields for investment. In March 1868 rich washings were found by sinking 15 feet through the surface soil on to the bed rock close to which the first pailful of dirt gave specimens, which I saw, consisting of four or five nuggets, one of which weighed 13 grains, a small piece of worn rock rich in gold and some dust.

Discoveries that from appearances at the time of reporting were likely to become of importance were noticed by the Commissioner as having been made between Old Tangier and Upper Musquodoboit; at Killag a branch of the Middle river of Sheet Harbour; at Fifteen Mile stream, a branch of the East river of Sheet Harbour, to which place a crusher has been transported during the winter; at Mosher's River; at Scraggy Lake, Ship Harbour; at Upper Stewiacke, where great excitement was caused last summer and many prospecting licenses and leases were applied for, and prospecting carried on with success; and at Gold River, Lunenburg.

From results of a visit paid in 1866, to the last named place, I reported that there was good promise of further working as the quartz veins were numerous and some had given from 16 to 22 dwt. gold to the ton. A new company began last year to work Mr. Sutherland's claims, on which I had reported. Quartz gave on assay in New York during the mining, 3 oz. 17 dwt. gold, and 12 oz. silver to the ton; the gossan, said to be abundant, gave me about 6 dwt. gold to the ton, and sand found later on gave me 14 dwts. 10 grs. retorted gold to the ton. A crusher was erected but, before it went into operation I believe, want of funds caused the breaking up of the company.

Since the issue of the Chief Commissioner's latest Report very considerable developments have been made in various districts and a large number of new companies have been incorporated. I refrain from giving any details as the accounts of operations are altogether too uncertain to warrant any statement not based on personal observation or official information.

GOLD.

The following are the statistical returns, in the Chief Commissioner's Report, for the year ending Sept. 30th, 1867.

Statement shewing the average daily labor employed, the amount of quartz crushed, the yield of gold per ton of quartz, the quantities of gold from alluvial mines, the yield of gold, the maximum yield per ton in each district, and in the whole province, and the value of the average yield of gold per man employed in mining, for twelve months ending Sept. 30, 1867.

DISTRICTS.	Average Men employed.	Crushing Mills employed Sept. 30, '67.	Steam Power.	Water Power.	Quartz, &c., Crushed.	Yield per Ton.	Gold from Alluvial Mines.	Total Yield of Gold.	Maximum Yield per Ton.	Average yield per Man for Twelve Months at $18.50 per oz.
					tons. cwt. lb.	oz. dwt. gr.	oz. dwt. gr.	oz. dwt. gr.	oz. dwt.gr	
Stormont, "Isaac's Harbor"	45	2	2	..	1149 00 00	1 05 08	....	1505 02 11	4 10 00	$618 73
Wine Harbor	33	4	3	1	1667 00 00	08 13	....	764 09 09	26 13 08	428 60
Sherbrooke	99	5	5	..	5809 00 00	1 09 08	....	8522 08 11	11 13 05	1592 58
Tangier	19	4	2	2	486 00 00	16 07	20 06 00	395 16 10	4 06 16	385 50
Montagu	19	1	1	..	214 00 00	1 19 00	....	417 13 21	2 09 20	406 60
Waverley	181	5	4	1	11289 00 00	07 07	....	4134 18 17	1 12 18	422 63
Oldham	52	4	3	1	960 00 00	1 08 07	....	1359 12 02	4 00 20	483 88
Renfrew	189	5	3	2	7770 00 00	1 04 04	....	9491 02 10	3 08 01	895 30
Uniacke	30	3	3	..	1212 00 00	15 15	....	947 01 17	14 10 00	584 00
Unproclaimed and other Districts	9	2	1	1	117 00 00	1 03 04	28 15 15	135 00 21	2 00 00	278 55
	676	35	27	8	36073 00 00	17 32	49 01 15	27583 06 09	26 13 08	$765 00

The following further statements and comparisons from official returns, Mr. Heatherington's *Guide*, Mr. Phillips' *Mining and Metallurgy of gold and silver*, and the *Report to Congress on the Mineral resources of the United States* by special commissioners Browne and Ross, will be found valuable.

The cost of raising a ton of quartz varies very much between the limits of $3 and $30; of 19 returns only three give $20 and above as a maximum; the average of the rest would be about $8. The average cost of crushing and mill treatment is 60 cents a ton by water and $2 by steam mills. In one steam mill the cost is given as one dollar per ton hauling included. In California the average cost in water mills, where the water is free is $1.22, where the water is bought $1.60; in steam mills $2.14. Miners' wages have varied little since gold mining began. Boys and carters have from 75 cents to $1; pitmen $1.10 to $1.50; mechanics $1.50 to $2.00 per day. In Australia the wages vary from $2 to $3.50; in California from $3.50 to $5.00, but it is reported that the miners have struck there for $6. Amalgamators have here $75 to $100 a month and mining captains $75 to $150 per month. Board and lodging can be had at most of the mines for $3 or $3.50 a week.

The sustained average yield of gold in the province from the crushing of 90850 tons of quartz was 1 oz. 0 dwt. 13 grains.

The annual quinquennial mean yield of gold per miner for the years ending 31st of December, 1862 to 1864, was $517.32; the triennial mean for 1864 to 1866, $680.90; and the biennial mean for 1865 to 1866, $744.16. The mean for 1867 was $765. (Year ending 30th Sep.)

The mean annual earning per miner in Victoria, Australia, was highest in 1866 when it amounted to $402.06 which is considered to be at least equal to the average earnings of the State of California.

The amount of quartz raised daily per man here has increased from 85.5 lb. in 1860-1861 to 300 lb. in 1866, a result shewing greatly improved system of mining.

The average yield of gold per ton of quartz in Nova Scotia compares very favourably with that of other countries. Mr. Ashburner, in 1866, reported that "it is very difficult to state, even approximately, the present average yield of the quartz from the Californian mines. It is probable however, that it has not varied much within the last five years and in 1861, taking the returns from those mines which were at that time believed to be profitable concerns, it was at the rate of $18.50 per ton. The two extremes were a mine in Grass Valley, which was yielding $80 per ton, and one at Angels, in Calaveras county, where the quartz only paid $5 and was still being worked at a small profit. The most noted of the excellent mines on the course of the "Great Vein" in Amador county, is the Eureka which has been worked for about 11 years and has produced probably nearly as much gold as any other in California. Its quartz has never averaged very high and the principal production has been from ores of a low grade, not yielding more probably than from $10 to $15 per ton." The average for the famous Brazilian mine St. John del Rey which, during 36 years, has given a profit of upwards of £1,000,000 sterling, is 8 dwt. of gold to the ton; during the most prosperous period of exploitation of the Berezovsk mines in the Urals, the yield was from 6 to 11 dwt., and at Zell, in the Tyrol, the average was only $2\frac{3}{4}$ dwt. of gold to the ton, being probably the smallest amount obtained from any rock worked. Victoria, Australia, gave for 1864 to 1866 an average of 10 dwt. $19 \frac{3}{10}$ grains, Nova Scotia for the same period, an average of 1 oz.

0 dwt. 13 grains of gold to the ton of 2240 lb.; the American ton, it will be remembered, is 2000 lb. and as Mr. Heatherington gives the average yield in California as 15 dwt. the two countries will probably be about equal in this respect. When it is recollected that in the provincial returns the statement is made with regard to the gold yielded by "quartz, sand, and gravel, crushed," although it is known that by far the largest amount of this is quartz, it is obvious that exact results cannot be obtained: it must also be taken into consideration that much gold is known to be stolen and much lost in working.

As regards the total production in California, although the rich shallow diggings are exhausted, by far the larger proportion, probably two thirds, of the gold is obtained from hydraulic and other "washings".

It may be well to record the following statement here for the sake of allowing comparisons to be made with what appears to be the extreme limit of productiveness. "Grass Valley in California is the most productive gold quartz mining district in the world. The annual yield of an area drawn by a radius of four miles is $3,500,000. The number of labourers employed in the mines and mills is 2000, showing an average yearly production for each person of $1,750 and the average yield of the rock worked is $30 to $35. The lodes are narrow, none of them exceeding 7 feet in width and most being less than a foot. They contain much pyrites and this fact with the narrowness of the veins contribute to make the average expense of extraction and reduction high—about $15 a ton. Some of the works have been sunk to a depth of 400 feet, but most of the pay quartz is obtained within 200 feet of the surface."

One of the provincial districts, Sherbrooke, it may be recalled, gave in 1866 an average yield per man of $1617.45; in 1867, of $1642.30. The success of the Ophir Company working at Renfrew having been mentioned as an instance of profitable working of quartz it will suffice to add that the Palmerston Company at Sherbrooke with 29 areas of which only 9 are worked had as the result of one year's operations $18,000 balance after meeting all current expenses, erecting a new 10 stamp mill, and paying a dividend of 25 per cent.

Mr. Rawson, writing, in February 1868, to the London Mining

Journal, said that three months inspection of the gold fields of the province had satisfied him as to their great value and made him anxious to arouse the capitalists of England to examine into their real importance. The American men of money he found had already learned their richness and were rapidly buying up all the best properties. Several of the mines he found quoted in the New York Exchange at premiums ranging from 120 to 150 per cent on their original shares.

M. Michel, who visited these mines last autumn and has been engaged in gold mining in South America, told me he was not at all prepared to find anything like the display of gold bearing rocks he witnessed and he thought that comparatively nothing had yet been done in mining towards developing the real value of the Nova Scotian auriferous deposits.

The gold of this province is not surpassed by any in the world in purity; repeated assays are said to have shewn it to average 22 carats fine, or to contain 916.66 parts of gold in a thousand; the official calculations are made on the assumed value of $18.50 for unsmelted or retorted, and $19.50 for smelted, gold. The actual composition of the gold as found native or in ingots or buttons is shewn in the following

ANALYSES OF NOVA SCOTIA GOLD.

Locality.	Authority.	Gold.	Silver	Iron.	Cop'r.	Lead.	Zinc.	Total.
Old Tangier	O. C. Marsh.	98.13	1.76	trace.	0.05			99.94
Locality not known	J. F. Baker.	97.30	2.70					100.00
Old Tangier (Field Lode)	B. Silliman.	97.25	2.75					100.00
" (Leary Lode).	U. S. Assay Office.	96.60						
"	A. Gesner.	96.50	2.00	0.05	0.08	0.06		98.69
Waverley (Laidlaw's)	H. How.	94.69	4.74		0.39		0.16	99.98
Ovens	A. Gesner.	93.06	6.60		0.09			99.75
"	O. C. Marsh.	92.04	7.76	trace.	0.11			99.91

The mean of the foregoing analyses gives 956.03 parts of gold in a thousand, the average of Australian alloy is given by Ashburner as 921, that of California by Dana as from 875 to 885 thousandths in gold.

The loss of weight in smelting the rough product of amalgamation is very variable, shewing that the retorted gold is of very uncertain value. In a great number of results obtained by Mr. R. G. Fraser, of Halifax, I found by calculation that the loss varied from 0.87 per cent on two samples from one district to **13.8 per cent on 13 samples from another district**, in two cases it was about

4½ per cent. on five and six samples, respectively, from other districts.

It is well known that a considerable amount of gold is lost by imperfect amalgamation : the following estimates have been made as to the actual amount in some cases.

Pyrites worked from tailings at the Nova Scotia and New York Co.'s mill at Tangier gave on assay by Dr. Torrey, Chief assayer to the U. S. Assay Office.

Gold per ton of 2000 lb................$122.13
Silver ,,   ,,   ,, ... 2.67
                                         ———————
                                          $124.80

Pyrites from Lake Co.'s lead, at Tangier, gave on assay by O. D. Allen,

Gold per ton of 2000 lb..................$187.04

Pyrites from tailings of the Leary Lode, Tangier, gave E. N. Kent

Gold per ton of 2000 lb..................$134.99

Assuming an average of 8 per cent as the amount of pyrites in the ore, and it has been variously said to contain from 8 to 12, the gold value of the pyrites will be $15.20 per ton. (Silliman's Report.)

Tailings from Waverley, selected from different parts of tailings bank :—

No. 1. gave 6 dwt. 8 grs. gold to the ton of 2000 lb.

No. 2. from five lb. of tailings all the silicious particles were washed out, leaving 3 oz. 11 dwt. of sulphurets, which gave 6 oz. 14 dwt. 1 gr. of gold, and 10 dwt. silver, to the ton of 2000 lb. On calculation this result gives $7.78 as the value of one ton of 2000 lb

No. 3. Tailings from the bed gave a value of $7.59 for the ton of 2000 lb., the value of the silver is not considered. (H. Perley.)

No knowledge appears to exist as to the actual amount lost in this way in California; the only data Phillips could collect on the subject were those obtained from a company in Australia, at whose establishment the tailings were regularly sampled and assayed ; the results showed that the tailings contained on an average about 2 dwt. of gold of which some portion was recovered by roasting and grinding the washed sulphides in a Chilian mill. The machinery and contrivances employed with the object of saving as much gold as possible are of the most varied characters and Mr. Phillips, finding it impossible to attempt a description of them all, confines himself to noticing such as are most commonly made use of in the best conducted establishments. These are described, and in some cases figured, in his Mining and Metallurgy of Gold and Silver, (p. 190 et seq.) and are well worth the attention of managers and companies. As all authentic data are of service in working out such a problem it will be useful to give some information obtained last summer while visiting Waverley. Mr. Burkner thinks the gold becomes mechanically attached to the pyrites in the process of stamping; the heavy pyrites remains in the stamping box a month, and he has seen on the ore under magnifying power distinct particles of gold, when, as he believes, there was not much gold originally in the pyrites. He sent to Freiberg, 1st, 850 lb. of pyrites from cleanings of stamper boxes in which quartz of 2½ to 3 oz. per ton had been crushed, the assay return was $600 to the ton; 2nd, two tons of pyrites from quartz of 15 to 20 dwt. per ton, assay return was $150 per ton; 3rd, 8 or 10 tons pyrites from tailings which gave $18 or $20 to the ton; the first and second lots of pyrites gave about as much quicksilver as gold, hence the mineral carries amalgam in close admixture.

CHAPTER IV.

SILVER.—ARGENTIFEROUS GALENA.—ANTIMONY.—MERCURY.—
MOLYBDENUM.—ARSENIC.—COBALT.—NICKEL.—BISMUTH.

Silver. Mackenzie River, Inverness Co., Cape Breton. It has long been supposed from Indian and local traditions that silver exists in the gravels of the bed of this river. Mr. John Campbell was the first to prove the fact of its occurrence: in his report on the Gold Fields, 1863, he states "that the prospects for silver mining appear most encouraging over a considerable tract of country, more particularly in the neighborhood of Grand Anse where the Mackenzie River falls into the Gulf of St. Lawrence. Native silver is found abundantly disseminated through the drift of this stream, in small grains and nuggets, and this appears to be the case along the greater part of its course, for in many trials made several miles inland I found the silver as plentiful as I found it near the Gulf coast. Nor is there reason to doubt the existence of rich deposits in some places where circumstances favoured the concentration of such particles of the drift as were of the greatest specific gravity such as silver. The sources from which this stream derived the silver rolled in its drift are, as far as I have been able to discover, first, from veins of a beautiful variety of spar, closely resembling meerschaum, that abound in some parts of the district; some of the veins contain native silver, embedded in strings and nests of a softish grey substance of earthy texture much resembling the carbonate of that metal. The other source I have reason to believe is the general surface of glacial drift along its banks and tributaries." Mr. Barnes has also found silver for upwards of eleven miles from the mouth of the river; he failed however, to discover the source of the metal.

Silver in the Gold Districts. Silver is contained perhaps invariably in native gold; it is shown in every one of the analyses of that metal, as found native, given; it appears also to exist separately. Mr. P. S. Hamilton, in his report as Commissioner of Mines, in 1865, states, "It would here be a needless discussion for me to enter into an explanation of the reasons for saying, what I nevertheless do say, that silver will probably be found in sufficient quantity to remunerate the miner in the same geological formation as gold. I believe I may safely say I have myself found it in some lodes now being worked." He then recommends such an alteration in the mining leases as would allow the miners to extract the silver. In Mr. Heatherington's Guide to the Gold Fields Mr. Pulsiver states that he knows where silver exists and only awaits an opportunity to avail himself of his knowledge.

Silver in Manganese Ore. Although it is not known in what form the metal occurs, it may conveniently be recorded here that Messrs. Taylor & Co. of London, found, on assay of a specimen of manganese ore from Teny Cape, Hants county, five ounces of silver to a ton of ore.

Silver in Coal. Mr. J. Campbell informs me that traces of silver have been detected in coal of Pictou county.

Silver in Native Copper. I have found a little silver in the copper of the trap region, as mentioned in speaking of that metal.

Silver Ores, Watchabuckt, Cape Breton. In the spring of 1865 Mr. A. Cameron, of Baddeck, brought me some specimens for examination one of which I found to be very rich in silver, the metallic ore was in fact sulphide of silver, it was in a quartz rock which also contained galena. Mr. Cameron afterwards had large samples assayed in Boston, and the property was surveyed by Prof. Hind, from whose report I make some extracts. "The Silver Mine.—This vein has been opened to a vertical depth of 15 feet near the junction of a bed of quartzite and coarse hornblendic rock with epidote, about $\frac{1}{3}$ of a mile from the south-west boundary of licence to search No. 9, on the peninsular south-west of Baddeck forming part of the south-west coast of the Little Bras d'or Lake, Victoria county. The vein occupies a fissure of variable width (from half an inch to four inches as far as traced) running nearly at right angles to the strike of the rock. Several much smaller veins run into it. The ore taken from this vein, assayed by Dr. Hayes, of Boston, gave these results: One ton 2000 lb. would afford 18 oz. 9 dwt. 3 grs. of gold, and 97 oz. 10 dwt. 4 grs. of silver, worth together $508.61 per ton of ore equal to the sample examined. From information obtained on the spot the argentiferous and auriferous ore appears to be present in the vein in pockets and not evenly distributed, hence it is probable that important aggregates of these metals may be reached. The angle at which the vein is inclined is low (27°), hence it is not probable that it will attain considerable width, yet if it maintains its present average thickness, notwithstanding its pockety character, the extraordinary richness of yield both of gold and silver, makes it valuable. The ore may be shipped in the capacious, safe, and deep harbour which lies at the foot of the hill about a third of a mile distant."

Argentiferous Galena. In the spring of 1865 I received several samples from Mr. Cameron some of which proved to contain silver also but not in such quantity as the preceding. The metallic minerals were galena, iron and copper pyrites, and a little blende some of which proved to contain a good deal of cadmium; gold was found in the quartzose matrix of the minerals. The report of Prof. Hind, in referring to the galena, says: "The argentiferous galena veins are situated in that part of the metamorphic nucleus which, in the form of high cliffs, constitutes part of the coast of the peninsula on Little Bras d'or. The rocks consist of alternating beds of quartzite, diorite, fine hornblendic slate, and coarse hornblendic schist. The main cliff vein is of uneven thickness, and, where exposed on the face of the cliff, about 65 feet above tide, it is from 12 to 16 inches thick. It is very ferruginous and contains thin seams of argentiferous galena with small pockets and specks of copper ore and iron pyrites. The analysis by Dr. Hayes of specimens from the cliff vein gave for the highest yield 39 oz. 10 dwt. 12 grs., or $51.38 worth of silver, to the ton. The argentiferous galena does not appear to be uniformly distributed through the vein, but is in patches and streaks. "In a conglomerate on the coast of the bay separating the peninsula from Baddeck argentiferous galena is found with copper ore in a band two feet thick.

In one sample of this Dr. Hayes found with gold, 6 dwt. 12 grs. silver to the ton."

For silver in galena of Gay's River see description of the latter.

Antimony. A small specimen of this metal, in the **native** state, was given me as coming from Halifax county, within a few miles of the city, without information as to the mode of occurrence or quantity found.

The existence of the rich antimony deposits consisting of the sulphide with occasionally a notable amount of the native metal, in the lower silurian rocks of New Brunswick, which are now being extensively worked renders this indication of importance. The uses of the metal are many and increasing: its most extensive application is in the making of valuable alloys. The ore of New Brunswick gives on assay from 47 to 73 per cent. of metal, and £14 sterling have been offered in England for certain qualities. A chemical manufacturer told me he had tons of this ore in New York where it is imported for use as a horse medicine; roasted to oxide it is employed in making opalescent glass, and converted into chloride it is used by calico printers. It is stated by Dr. Hayes that the ore costs about $60 a ton imported to Boston. (See reports on Geology of New Brunswick by Prof. Bailey, 1864, and Prof. Hind, 1865.)

Mercury. This metal and its sulphide, cinnabar, are said to have been found at Gay's River, Colchester County. Professor Lawson informed me that he had good reason to know that mercury was formerly found in the province. Mr. Burkner showed me a shaft at Waverley, from which he had taken a quantity of soft slate affording him by washing, on repeated trials with clean pans, metallic mercury: he believed cinnabar was also present.

Molybdenum. The ore of this metal known as molybdenite, or the sulphide of molybdenum, is found at a few places. The most abundant source, so far, has proved to be quartz at Gabarus, Cape Breton; the mineral is sometimes well crystallized: I have seen a few ounces from this locality. It is also found at Hammond's Plains and Musquodoboit, Halifax County, and at Chester, Lunenburg county. It is not put to any use in large quantity; it is employed in a very limited application. It is known to afford fine colours but has never been met with in such quantities as to encourage manufacturers to experiment on an expensive scale. I remember some three shillings a pound being offered for it in London.

Arsenic. The large quantities of mispickel found in some parts of the province as an associate of gold might be made to furnish the arsenical compounds of commerce. The mineral is composed of 19.6 sulphur, 34.4 iron, and 46 per cent. metallic arsenic: when roasted it gives the white arsenic of the shops and this oxidised furnishes the arsenic acid so largely used in producing some of the beautiful new colours from coal tar. For this latter use one manufacturer ordered on one occasion nearly 100 tons of the acid. The total amount of arsenical compounds used must be very large: in 1866 a company was formed in England whose works had produced 500 tons of white arsenic annually and it was proposed to extend them and to utilize the mineral residuum of the arsenic sublimation process. This residue would contain in the case of some of the mispickel of this province both gold and cobalt—perhaps nickel also. The total quantity of crude arsenic produced in Devon and Cornwall in 1861 was about 1210 tons, of the value of about £1126, total fine arsenic being 1450 tons, value £10,785 sterling.

Cobalt and Nickel. Not only is cobalt found as just mentioned in mispickel; (according to reports from Freiberg it is sometimes present in considerable quantity in that from Montague,) but it occurs in small amount in other minerals. I have found it along with nickel, together amounting to two thousandths estimated as oxides, in magnesia alum (as shewn in the analysis given) from Newport. I examined the rock in which the mineral was found for the metals but found neither of them, this however is no proof that they do not exist in other parts of the rock because they probably occur as sulphides and the action which allowed of their being in the alum to which they are altogether foreign is exactly that which would remove them from the rock: the specimens I tried were from the edge of the rock where the alum had effloresced, it is quite probable that an examination of the unaltered rock within would show different results, and possibly the metals may be found in the rock in remunerative amount. I detected also about the same quantity of these metals in magnetic iron pyrites from Nictau, Annapolis Co., in that from Geyser's Hill, Halifax Co., and distinct evidence of their presence, perhaps to the same extent, in a rock from Cape Breton. I found cobalt in wad from one locality, and no doubt, as it is often met with in this mineral, it exists in various places where this is abundant. Dr. Dawson gives 2.10 per cent as the amount of cobalt in copper ore from Carribou, Pictou Co.

With regard to these metals a small percentage pays ; magnetic iron ore with 4 per cent of nickel has long been worked in Italy, and I was told by a gentleman returning to England from Newfoundland, where he had been in search of metals, that the ore from which most of the nickel of commerce is obtained contains about 2 per cent on the average. The chief use of cobalt is in making the well known blue colour seen in glass, pottery, and paper, and of nickel in making alloys. German silver or nickel silver contains copper 8 parts, nickel 3 parts, and zinc $3\frac{1}{2}$ parts ; the small cent pieces of the United States and some small coins of Switzerland consist of copper and nickel. The ores of these metals are mostly obtained on the Continent of Europe. In 1860 the total produce of British ore was less than 7 tons, in 1855 about 350 tons had been raised. Nickeliferous ores of course will vary much in value, some are stated to be worth about £30 sterling a ton.

Bismuth. Several years ago when in Cape Breton I was informed that this metal had been found native there, and lately Mr. Barnes has told me that he met with it in water-worn nuggets from the size of a wheat grain to that of a pigeon's egg, at Wagamatcook, Inverness Co. So far as I recollect I was told on the first occasion that the metal had been got in the rocks by a man who could not find the locality on a second search ; Mr. Barnes found it in a river-bed. The price of this metal has been of late years most curiously subject to variation owing to the formation of a company in London for the purpose of making gold by transmutation of base metals. Bismuth was to have been largely used and all that could be got was purchased regardless of price. Of course no gold was made and, to save the utmost out of the wreck, the deluded shareholders tried to sell their stock of bismuth so as to obtain as high a price as they could, and thus, by a process at least possible, to convert it into gold. In 1844, bismuth was 10d. a pound, in 1844 and 1845 2s. 6d. and this price continued till 1858, except on one or two occasions when it rose to 4s. per pound. In 1861 it suddenly advanced to 9s. 6d., and it reached 20 shillings in 1862, since which time it has fallen back to 10s. 6d. The principal uses to which this metal is applied are in the manufacture of fusible metal, of calico printers' rollers, and of medicines. It and its ores are not found in great abundance though in 1865 it was stated that the metal was being made in large quantities in the Iljampu mountain, in Chili. England is chiefly supplied from Saxony and Bohemia: although found in Cornwall the metal is but rarely obtained in sufficient quantities for sale. (Qu. Jnl. Science, 1864 and 1865.)

CHAPTER V.

COPPER.—COPPER ORES.—LEAD ORE.—ZINC ORE.—PLUMBAGO.— SULPHUR—SULPHUR ORE.

Copper. This metal is found at several localities on the shores of the Bay of Fundy and of Minas Basin. The name Cape d'Or is supposed to have been given by the early French mariners to the mass of trap forming the south-western extremity of the county of Cumberland from their mistaking the pieces of metal seen in the rock for gold. The metal is found in rounded and flattened pieces imbedded in the trap and in similar forms in the shingle at its base; a specimen of 15 lb. weight is reported to have been obtained on the west side of the Cape. It is said to be always more abundant after a storm, either from disintegration of the rock or from being washed up on the beach. It is found on the east side of the Cape in a vein of jasper, and on the surface in the soil. A specimen was lately given to me consisting of the metal inclosed in red copper ore and green carbonate of copper: the whole weighed very nearly a pound. Various statements are made as to the amount of copper existing here, a few years ago little could be found, while it is said by the residents in the neighbourhood that in 1866 about two tons were carried away by Americans. Specimens quite similar in appearance to those from the Cape are obtained at Spencer's Island a few miles to the east. Some 20 miles further east the metal is again found in trap and is reported to have been got in holes at low tide in this vicinity to the amount of a few pounds; specimens have been shewn me as being so found. Last summer two fine specimens were submitted to me, said to have been got on the shore of Cumberland county, one was in cubic crystals, the other a rounded flattened mass, their weight was perhaps half a pound. Mr. R. G. Fraser told me he once had a piece of copper from the

Basin of Mines weighing six pounds. In the trap rock forming the southern shore of the Bay of Fundy copper has been found at various points from Cape Blomidon, at the eastern extremity, to Brier Island at the western extremity, of the coast.

The most productive locality has been thought to be Margaretville or Peter's Point. I saw it in this neighbourhood 12 years ago. In 1862 a company, formed in Halifax, prospected here and commenced operations at Bishop's Brook some 2 or 3 miles east of the Village. The metal was found crystallized in a matrix of zeolite running through the trap, and appeared to be most abundant in the rock on the beach exposed at half-tide. A considerable excavation was made and about three hundred weight of metal got out during the summer. I visited the spot the next summer and found the place abandoned. The metal occurs close to the wharf at Margaretville in zeolite in the trap and also in rocks exposed at half-tide, as well as near a brook a few yards to the east and again about 1½ mile back from the shore. An American company has been working at this last locality; in 1864 $2500 were expended in operations: the licensees appeared sanguine as to their eventual success. Operations however have been abandoned.

The enormous deposits of copper at Lake Superior are, as at the foregoing localities, found in trap, and sometimes also the metal is associated with zeolites. Whitney, however, has shewn that though the rock there contains a few of the zeolites met with here, those minerals which are most characteristic of the Nova Scotian trap rocks are almost entirely wanting at the Lake, and that while no analysis of the rocks themselves has shewn their geological age to be identical, their mineral contents fail entirely to indicate this being the case. (Silliman's Journal, July 1859, p. 20.) I find the copper from Cape d'Or to yield on analysis a trace of silver of which a specimen from Lake Superior, apparently pure, contained according to Whitney three ten thousandths. Silver is not visible in the Nova Scotian copper as it very often is in considerable masses in that of Lake Superior. Mr. Barnes informs me that native copper has been found in the trap of Mt. Jerome, 4 miles north of Cheticamp, Cape Breton.

Although there may be difference in the geological age of the trap here and at Lake Superior, the occurrence of copper at many places in the Bay of Fundy and apparently the Basin of Minas is important and in this connection it is worth observation that at the Lake Portage copper district of Lake Superior where there are at least 12 mines in operation of which the majority are producing from 20 to 120 tons pure metal per month, some of the works on the most productive lode make use of rocks containing quite a low per centage of native copper. Thus at the Quincy mine, the width of the copper bearing rock being from 6 to 30 feet, the average being 10 feet, the amount of metal was on calculation from the yield per cubic fathom in 1864 only 1.4 per cent. At the Pewabic and Franklin stamp works the yield of the rock treated was 1.69 per cent. Of course the copper being unequally distributed through the rock the true percentage would often be above, often below, these numbers. The metal is got by burning the rock, and subsequent stamping and washing with magnificent machinery. (Geology of Canada, 1866, pp. 162-3.)

Ores of Copper. These are found at many localities and are occasionally very rich: the quantity in some places being such as to have led to the commencement of mining operations.

Copper Pyrites at Polson's Lake, Antigonish County. On the south side of Polson's Lake, about two miles east of Lochaber Lake large fragments of copper and iron pyrites in an impure brown hematite are found in the surface gravel, sometimes these are from three to five feet in diameter; the average amount of copper of several samples was found by Dr. Dawson to be 10.8 per cent. From Dr. Honeyman's survey of this district in 1864, for the Government, it appears that near the junction of the basal greenstone rocks of the series, at Keppoch, with the sedimentary rocks there are traces of carbonate of copper and numerous veins of micaceous iron ore. In the sedimentary rocks are small quantities of carbonate and sulphide of copper and in some of the devonian are the blue slates of Polson's Lake with numerous veins of carbonate of iron or ankerite with sulphide of copper, these veins being undoubtedly a continuation of the veins which produced the masses of cupriferous oxide of iron which have long attracted attention. The ore and gangue as found at Grant's Brook were sent to me for examination; the gangue proved to be carbonate of iron with a little carbonate of magnesia and perhaps one or two per cent. of carbonate of lime with a small amount of quartz and perhaps steatite. The ore was pyrites rich in copper. I afterwards had to examine other specimens from Lochaber, these consisted of iron and copper pyrites in iron ore of similar character to the preceding. The copper contained I found to be about 9 per cent.

Dr. Honeyman considered that there could be no reasonable doubt as to the position of the vein of copper ore at Polson's Lake. Leads were subsequently found and last year (1867) a company was incorporated for the purpose of working the ore and operations have been carried on to some extent and are still being pursued. If the ore proves really abundant and the average yield of copper remains as above indicated the prospects of the company are excellent. The improvement in working copper ores in England have made a very much lower and decreasing amount of metal remunerative. The following data from Phillips and Darlington's Records of Mining and Metallurgy are very instructive on this point :—

Average produce of Copper from the mines of Devon and Cornwall.

1776 to 1785 average produce 12 per cent. metal
1826 to 1835 ,, ,, 8 ,, ,,
1836 to 1845 ,, ,, 7½ ,, ,,
1846 to 1855 ,, ,, 7½ ,, ,,

"These figures" it is remarked "clearly shew the economic value of the improvements effected in working our mines."

Proof of the increasing excellence of the methods now in use is afforded by the corresponding details up to nearly the present time which I have taken from the Mineral Statistics of the United Kingdom shewing that the average produce from 1859 to 1866 varied from 6.65 in the former year up to 6.77, the highest reached, in 1863, and down to 6.18 the lowest percentage, given in the latest year named ; the price of the ore at this percentage was £4. 11s. sterling per ton. I may add that Phillips and Darlington are careful to point out that the mean produce of the vein stuff of Devon and Cornwall has been estimated by eminent authority at about 2 per cent. and that much of the exceedingly large amount of ore got at the Devon Great Consols affords not more than from ¾ to 1½ per cent. of metallic copper. Very poor ores are found to be remunerative on the continent of Europe. Thus in 1861 no less than 50,-

000 tons of coppery iron pyrites were exported from Portugal to England which contained at most 3½ to 4 per cent. copper. In Prussia the Stadtberger Company of Altena treat successfully ores containing only from ⅔ to 2 per cent. metal. Of the 65000 tons worked up every year, 12 parts out of 13 form what they call second class ores and contain only two thirds per cent. copper.

Ores of Tatamagouche, Cumberland Co. For the account of these I am indebted to Hon. A. Patterson. of Tatamagouche, who kindly procured a report from Mr. P. Brodie (at the time, 1866, in charge of the mining operations,) which is essentially as follows:

The place now being worked is on the north west side of French River, (about 5 miles from the village of Tatamagouche) at a point where the bank is about 70 feet high. After removing about 12 feet of surface I first came to a deposit of blue clay which contains some small nodules of copper ore, (grey sulphuret). This was about one foot thick, next came a layer of brownish grey sandstone about two feet thick, in the centre of which I find a good many nodules embedded of the largest size we get. I should properly divide this last for there is a regular wall-like separation just where the copper is most abundant. Next in order is a bed of clay similar to the first, which resolves itself into a sandy formation carrying copper as it gets deeper for about 18 inches, under which is the next productive stratum which is a dark grey sandstone completely charged with copper. This bed is a little over a foot thick and the ore is dispersed through it in nodules, although there is a great deal of ore lost for want of proper appliances. I should mention this formation of ground has a slight dip to the N. E. of about 1 in 12. On the south east side of the river I have driven a level about 20 feet on a nine inch lode of ore. It is a mixture of black and grey ore. This place differs from the first named inasmuch as there is no clay on either side, the lode being bound on top and bottom by good solid grey sandstone. The ground on this side the river rises to a gradual slope from the water, but on the north side the bank is abrupt. About the bank on the north side is another small irregular seam of copper, I have not opened on it, so can say nothing about it. Close down on the water's edge there is the outcrop of a lode of ore imbedded in a heavy stratum of conglomerate.

This lode is a very nice mixture of copper and lignite, it is about two inches thick at the point seen, but I am informed on reliable authority it is from six to eight inches where seen in the river bed. If it is as large as that, there is no doubt of its being a valuable mine to work. On the north east side where I began to work I have driven in 4 tunnels about 40 feet each, and have cross cut some of them. I find it very easy ground to work but it wants care and good timbering to make it secure. I sank a shaft about 150 feet from the edge of the bank and at the depth of 25 feet came to a small seam of ore. I have another shaft sunk to cut the lode containing the lignite but have not yet reached it. Altogether a good deal of work has been done this summer. I have six men nearly all the time at work. I do think if this place were properly tested it would turn out a good paying property; every appearance is in its favour. The yield of copper from samples sent to Boston is 74 per cent."

Specimens from both the first mentioned spots were sent to the Paris Exhibition. The samples from the south side of the river were sandstones containing numerous nodules of ore about the average size of bantam's eggs. The others were loose nodules of the same character and perhaps on the average larger. Some nodules found here are larger than hen's eggs. I found the specific gravity of a specimen from this locality to be 5.257, this and the great richness, 74 per cent of copper having been found on assay in Boston, show the ore to be vitreous copper; it is often coated with green carbonate.

Ores at Cheticamp, Cape Breton. These have led to the formation of a company which has carried on operations for a short time only. The official report on the mine in 1864 is this. "This mine is only in progress of development—all that can at present be said about it is that the indications are good. The vein rock was discovered at several points along a line in the directions of S. 40° W. about a mile and a half from the shore of Cheticamp. The vein rock is three feet six inches thick, and the vein itself about five inches. It dips towards the mountain at an angle of 60° with the horizon, in the direction S. 50° E. This mine was visited early in October; at that time a shaft was opened in the hard rock to a depth of 10 feet, and a small house, used as a smithery, erected over it. An adit or tunnel was also in course of construction. The adit was driven into the face of the hill about 100 feet, but was not expected to meet the vein until it had been driven 410 feet, at which point it was expected to intersect the vein rock, 106 feet from the surface." From information kindly furnished this year by Mr. Barnes, who superintended operations for some time, it appears that the vein rock at the surface is calcite which runs wedge-shaped between sandstone and micaceous slates, and that the ore at the outcrop consists of silicate of copper (chrysocolla) with blue and green carbonates, and spots of rich grey and yellow ore, while the veins intersected by the level are several in number and consist of rich yellow ore. On continuing the levels beyond the shaft into the mountain intrusive dykes of syenite containing flesh red felspar and bluish quartz were met with. There has been nothing done at the mine for two or three years, and I believe there is no present intention of resuming operations.

Ores on East River of Pictou. Promising appearances have been found by Mr. Barnes on the East River, about 4 miles from New Glasgow, near the railway. The ores consist of sulphuret with lignite and green carbonate disseminated through micaceous sandstones; the cupriferous bed is estimated to be about 2 feet thick and has been found on both sides of the river. I visited the spot last summer and saw the outcrop of the bed on one side of the river. Several specimens of the ore and rock subsequently sent me look very encouraging. Some of the vitreous ore with lignite gave Mr. Barnes nearly 40 per cent of copper. The rocks here are probably at the base of the coal measures or lower carboniferous, and an interesting case of copper ores being worked in lower coal formation sandstones occurred at Newgate in Connecticut. Here, Mr. Barnes informs me from observations of his own, and inquiries made at the time of his inspection, the sandstones resting on highly metamorphosed conglomerate contain a bed from eighteen inches to two feet in thickness of rich vitreous copper cased with chrysocolla. The rock was estimated to contain about four per cent. of ore, the dressed ore gave 37 per cent. of metal. Workings had been carried on by the State for about sixty years over fourteen acres and a very large amount of ore must have been taken out. The district was laid down as altered carboniferous, not lower carboniferous. This would appear to be an instance where copper has been profitably worked in rocks of this age. Mr. Barnes thought the appearances not so favourable as at East River.

Ores at Indian Point, Five Islands, Cumberland County. These consist of red oxide, with a little grey ore and green carbonate, in association with magnetic oxide of iron. They occur in veins from 2 to 10 inches thick in trap, the immediate matrix or vein stone being a hard jaspideous rock. Mining operations have been conducted here by Mr. J. Browne who shipped about 10 tons to England which gave some three per cent. of metal, a second quantity of about 12 tons, of richer appearance, was lost on the voyage. Some tons remain at the spot. The ore sent to England had not been in any way dressed, as, the veins being small, it was quarried out rock and all and so sent. Mr. Browne thinks that it might have been brought up to 10 per cent. by proper dressing. The ground presents great facilities for trying the deposit in depth, and a tunnel 300 feet deep it is thought could be brought in to strike the mineral-bearing ground. A specimen of the iron ore found gave me 5.6 per cent. metallic copper.

Ores in other localities. Very rich copper pyrites, yielding 31.6 per cent. of metal has been found on the south branch of the Salmon River, but Dr. Dawson who obtained this result is not aware of the ore occurring in quantity sufficient for mining purposes. This vein is probably the Salmon River of Guysboro' as I have had rich copper pyrites given me as coming from this county. Of the ores of the coal formation Dr. Dawson says (Acadian Geology, p. 267), "the principal localities are the Carriboo River, West River, a little below Durham, and East River, a few miles above the Albion Mines. Similar appearances also occur at French and Waugh's Rivers in the band of coal formation rocks connecting Cumberland and Pictou districts. In all these places the principal ore is grey sulphuret with films and coatings of green carbonate. The ores are associated with fossil plants to which their accumulation is to be attributed. The ores are rich and valuable and the only reason they are not worked is the conviction that the deposits are too limited to be of economic importance. This has been found to be the case in two instances in which trials have been made by agents of the Mining

Association. The following is the composition of a sample from Carriboo, copper 40, iron 11.06, cobalt 2.10, manganese 0.50, sulphur 25.42, carbonate of lime 0.92, total 80."

A very rich sample of copper ore was given me a few years ago as coming from Cumberland county, and one as being found at Margaretville, Annapolis Co. Specimens, of which I do not know the localities, sent me for analysis have given, among other low results, 40.25 and 39.13 per cent of copper, the latter contained in all:

```
Copper..............................................39.13
Iron................................................18.40
Sulphur.............................................11.90
Gangue..............................................18.58
Oxygen, carbonic acid, water, a little lime and loss..11.99
                                                    ──────
                                                    100.00
```

So long as ores containing so highly remunerative an amount of copper continue to be met with there will be very strong inducement to prove whether they exist in quantity sufficient to warrant mining operations, even in the face of repeated disappointment. Mr. J. Campbell, in his report on the Eastern Gold Fields, 1863, states that from the mouth of Steep Mountain river, Inverness Co., Cape Breton, for a distance of thirty miles eastward favourable indications of copper ores exist. Mr. Poole, in his report on the Western Gold Fields, 1862, mentions that at Blandford Cove, at the base of Aspatogoen Mt., Lunenburg Co., bands of dark blue ironstone slate visible for some distance along the shore hold a good deal of copper pyrites, and that it might be worth while to search for a lode, also that on Hillsborough Brook about 12 miles from Brookfield, in Liverpool township of the same county, a good deal of copper and iron pyrites was seen in quartz veins on which excavations had been made and which were said to increase to the east, and that copper might be found in depth. He observed copper pyrites also at the Westfield Brook in the same region, and in an official examination of some of the specimens he brought home I found a little copper in pyrites from Geyser's Hill, Halifax, and from Jebogue Point, Yarmouth, a circumstance leading Mr. Poole to remark that though small the amount given might induce parties to explore in depth as copper is not usually a surface metal. Some copper was said to have been dug out of cellars at Middle Jebogue many years before Mr. Poole's survey.

At the Provincial Exhibition now open (Oct. 1868) Mr. Hugh M'Adam shews specimens of promising copper ore from a deposit reported to be a thick bed in Antigonish county.

Lead Ore. The only ore of lead found in the province is galena, it occurs in many localities, but so far only one of these has been thought sufficiently rich to encourage mining operations. The ore is very frequently found with gold in quantities sufficient to be very troublesome in amalgamating, and occasionally it occurs with rich silver ores or itself contains a large amount of silver.

Galena of Gay's River, Colchester Co. This is the ore which has been thought to occur in quantity a favourable report having been made to this effect by a mining engineer, since deceased. The ore occurs in disseminated crystals and thin veins in beds of lower carboniferous limestone, and contains silver in varying proportions, as shown by the accompanying reports of assay masters. As regards quantity and mode of occurrence the late Mr. S. Bawden reported as follows:—

" R. Smith, Esq.,   Halifax, Sept. 27, 1862.

"Sir,—Having had samples of lead ore brought me to examine which I considered most valuable both for lead and silver, but at the same time, doubting in my mind if it could be had in such quantity as reported to me, and the owners of the property confessed to be totally ignorant as to its worth or how to work it, or how to dress the ore to make it marketable—this induced me to go to see it before leaving this country.

" At the first sight I was particularly struck with the quantity on the surface in the shape of boulders and flagstones lying one on the top of the other,—*not* a conglomerate, but solid massive rocks of the mineral as I will herein explain.

" A trench has been dug for some little depth by the proprietors, and loose earth been found deposited between those layers impregnated with the mineral of the richest quality.

" The main bottom has not been reached in the line of mineral, but on either side, some few yards distance, the hard whinrock is cropping out at the surface, and those vertical beds of mineral are deposited, as it were, in a hollow or bend in a gentle rise of the hill between the above mentioned whinrock.

" The result of my observation on this is that it is an upheaval from an immense deposit of mineral which will be found in the shape of a lode running east and west.

" What is more particularly noticeable is its formation of slaty cleavage, and the peculiar blue grain of strata found mixed in with the mineral ores, which is the evidence to show that these impressions were produced from a lode underneath, as there are no such indications on either side of its immediate course.

"The assay sent you from Mr. Nash, was what it was worth in its rough state, but the sample sent to Cornwall was first cleaned by the aid of water, and the ore then is worth £18 per ton, and eleven and a half ounces of silver to the ton.

"You need not however, take either of these, but when I return to England I will bring a fair average of the ores so that you can get assays in London.

"I visited the place again last week, and was as well pleased with its appearance as I was at first sight. Thousands of tons can be raised for 2s. per ton before the lode will be reached.

"I should not omit stating that this is close along side a river called Gay's River, where abundance of water can be had at all times for the purpose of cleaning and dressing the ore.

"All other information you deem necessary, I shall be able to give on my return. I am, Sir,
"Your most obedient servant,
"SAMUEL BAWDEN."

The following reports refer to the quality of the ore.

ASSAY OFFICE, 77, 78 & 79, HATTON GARDENS.
London, 21st Feby., 1862.

"The sample of mineral assayed for Messrs. Pixley, Abill & Langley is found to produce 8¾ per cent of good pig lead.

"The proportion of silver is equal to .35 oz. to the ton of 20 cwt. of ore. It contains minute traces of gold.

"As the lead in this mineral exists almost entirely in the state of galena, it may easily be treated by washing, thus reducing the ore in bulk by 80 or 85 per cent. and bringing it to a valuable and marketable condition.

"JOHNSON, MATTHEY & Co."

The quantities of metals given above, otherwise stated, are equal to 196 lb. of lead to the ton (English) of ore, and 4 oz. silver to the ton of lead. The next report is that of Mr. Rickard, made for Mr. J. D. Nash:

"Halifax, N. S., Dec. 31st, 1862.

"First. Eight oz. No. 1 galena ore gave on washing 1¼ oz. concentrated ore, yielding 27 per cent. lead and 44 per cent. rough ore, or 19 dwt. 11 gr. silver per ton of washed ore, or 3 oz. 11 dwt. 13 grs. silver per ton of lead.

"Second. No. 1. galena, Old Workings, 6 per cent. lead, 4 dwt. 21 grs. silver per ton ore.

"No. 2. On Hill, 8 per cent. lead, 6 dwt. 12 grs. silver per ton ore.
"No. 3. At Brook, 5 per cent. lead, 3 dwt. 6 grs. silver per ton ore.
"No. 1. F. 18 per cent lead, 16 dwt. 8 grs. silver per ton ore. Trace of gold.
"W. T. RICKARD, F. C. S."

It will be observed that the first assay by Mr. Rickard gives not very much less silver to the ton of lead, than that of Johnson & Co., the percentage of lead in the concentrated ore is equal to 604 lb. to the English ton. The last assay in the report, otherwise stated, gives 403 lb. lead to the ton, and $4\frac{1}{2}$ oz. silver to the ton of lead. The next report is from England,

"FLOCK LEAD SMELTING WORKS,
"Devon, Cornwall, Aug. 15th, 1866.
"Sample of lead ore from Nova Scotia, per James Richard Bawden.
"Lead $17\frac{3}{4}$ per cent
"Silver $1\frac{1}{2}$ oz per ton of ore, $11\frac{1}{2}$ oz. per ton of lead.
"H. B. CHIPMAN."

The ore contains, as found by Baron Liebig, who examined it for Mr. R. G. Fraser, antimony to a very small extent, a specimen afforded me also a trace of this metal.

With regard to the value of the silver in lead ores the following abridged extract from Phillips's Metallurgy (now out of print) is instructive. "Previous to the discovery of the present methods of improving and enriching the metal obtained directly from the ores none but moderately rich leads could be treated for the silver they contained. The method of treatment not only involved the expenditure of a large amount of coal, but also the loss of at least 7 per cent. of the lead operated on; consequently lead that did not contain from 9 to 11 oz. of silver to the ton, did not admit of being profitably refined. When the lead and silver were also associated with tin or antimony, the difficulty and expense of this process were much increased, and proportionally richer ores were consequently required. By the improved processes lead containing but three ounces of silver to the ton of metal may be refined with advantage." This was written in 1854 since which time no doubt the methods have been further improved. In 1864 the whole amount of lead ore raised in Great Britain and Ireland was 94,433 tons and from this there were obtained 67,081 tons of lead and 641,088 ounces of silver, the value of the lead being £1,448,959 and of the silver £176,299. In 1866 there were raised 91,047 tons of lead ore which gave 67,390 tons of lead and 636,188 ounces of silver; the percentage of lead in the ore is rather above 74 and the amount of silver in the ton of ore just under 7 ounces. The mean price of the lead ores raised in 1861 was £12 10s. 7d. Pig lead is worth about £21 a ton, silver about 5 shillings the ounce.

Argentiferous Galena of Baddeck, Cape Breton. See *Silver Ores.*

A specimen of galena of which the locality is said by Mr. W. M. Harrington, who sent it me for analysis, to be Lunenburg County gave me:—

    Lead.................................85.05
    Antimony ....traces.
    Sulphur and Gangue..................14.95
                                             100.000

No silver was detected in the wet way by a special test: minute traces may have been present—indeed it is said no galena is quite free of silver. The ore was very pure but proved not to be accessible or probably abundant. I have found several specimens from gold regions of the province to contain notable amounts of silver.

Galena of other localities. This ore is found in fine crystals at the Joggins Coal Mine, Cumberland county. Two very good specimens were brought me by a miner in 1858 as having been obtained about 6 fathoms below a seam of coal in cutting through a fault, the agent of the General Mining Association concluded there was not sufficient to be of importance. Galena is reported to have been found in small quantity in Guysboro. Mr. Barnes informs me that numerous veins containing much galena occur in altered limestones at the mouth of M'Kenzie River, Inverness county, Cape Breton; and that one of the veins contains large octohedral crystals of fluor spar of which some are upwards of an inch across. Mr. Campbell, in his report before referred to, also notices that masses of galena are found distributed through the tranverse sections of some large mineral veins seen in the sea cliffs between Fish Pond River and the mouth of the Mackenzie River. Though the ore is not in large quantity so far as can be seen on the surface, the facilities for mining are so favourable that comparatively poor ores might be profitably worked. It is worthy of notice, as Dr. Dawson points out, that the lower carboniferous limestones which are so abundant in this province are the rocks in which valuable ores of lead are met with in England and other countries, and that therefore there is some reason to hope that important indications of this metal may yet be discovered.

Zinc Ore. The only ore of zinc yet met with here is the sulphuret, called blende and, when of a dark colour, black-jack. It is seen very frequently in association with gold in quartz, but the quantity has not been found to be of mining importance. The largest pieces of it I have seen were from Mount Uniacke, they were crystalline masses, perhaps half an inch across, of a very dark colour, containing a good deal of iron: they were associated with cale-spar and a little pyrites in quartz.

Tin Ore. Tinstone has been found by Mr. Barnes in a sand composed of quartz and decomposed felspar in Tangier and by Mr. J. Campbell at Shelburne as I have understood.

Alloy of copper, zinc, and tin, reported to have been found in the province. I place on record here the analysis of a substance of metallic appearance and somewhat tin-like colour on fresh surfaces sent to me some years ago and which was said at first to be abundant but afterwards only to occur in thin veins. The composition found being so entirely unlike that of any known mineral and no matrix being furnished with the specimens examined by me mineralogists (specimens were sent to the London Exhibition of 1862) have refused to believe in the substance being a mineral. However, the person from whom I received it on being applied to for information asserted that it occurred in veins of which he gave the direction. Specimens afforded me on analysis:—

    Copper 57.29
    Zinc 20.54
    Tin 19.94
    Lead 2.39
    Iron 0.08
                                               ————
                                               100.24

While there is no mineral having a composition similar to or approaching that shewn by these results, there is also, so far as I have been able to find, no artificial alloy having any thing like the same proportions of ingredients. A mining engineer from England to whom I showed specimens offered £75 sterling a ton for the substance of the quality shewn by my analysis.

Plumbago. Specimens of earthy plumbago or black lead have been sent to me for examination from Parrsboro', Cumberland county, where it occurs at Partridge Island, either in a bed or vein on the shore ; and from Salmon River, Colchester county. The former was in powder, the latter in masses some inches thick. Mr. Fraser informs me that a deposit of 10 inches thickness occurs in Cape Breton, and one of 4 feet in Musquodoboit, both of an earthy nature. The same gentleman has lately shewn me the best quality of plumbago I have yet seen as found in the province. It is said to be in a deposit six feet thick in a new locality. These deposits may be worth attention since the improvements in the preparation have rendered impure plumbago valuable. The chief deposit of pure plumbago was at Borrowdale, in Cumberland, England, where it was found in detached pieces very irregularly distributed. The supply was uncertain and of doubtful continuance. The price of the mineral was kept up by careful management; about 15 years ago it was 35 to 45 shillings a pound; and at this rate the mine has been known to bring £100,000 in one year; in 1859, the price was 30 shillings a pound. The present price of plumbago from Ceylon is, for lump, 25s. to 39s.; for dust, 13s. to 20s.; from Germany, 7s. to 10s. per cwt. The use of the finest mineral is in making pencils : the best of these are made by sawing the plumbago into pieces which are inserted in cedar, inferior qualities are made of the saw dust and small pieces mixed with antimony and sometimes sulphur. A considerable portion of the smaller pieces of the Cumberland plumbago which are too small to cut for pencils is of the finest quality. Mr. Brockenden patented a process by which these were ground, and ultimately formed by enormous pressure into coherent blocks from which slices could be sawn fit for making pencils of the best quality. Another process, invented by Brodie, removes from common plumbago by chemical means the impurities which prevent the formation of solid blocks by Brockenden's patent.

The mineral in coarse powder is submitted in an iron vessel to the action of twice its weight of oil of vitriol and seven per cent of chlorate of potash: the mixture is heated over a water bath till the chlorous fumes cease to be given off. By this means the compounds of iron, lime, and alumina present are rendered for the most part soluble and the subsequent addition of a little fluoride of sodium to the acid mixture will decompose any silicates which may remain and volatilize the silica. The mass is now washed with water, dried, and heated to redness. The last operation causes the grains of plumbago to exfoliate, the mass swells up in a surprising manner and is reduced to a state of very fine division. It is then levigated and obtained in a pure condition ready to be compressed by the method of Brockenden before mentioned by which coherent masses are obtained equal in beauty and solidity to the best native mineral.

Very large quantities of plumbago are employed in making, beside lead pencils, glazing for gunpowder, polishing powder for stoves, paints, and crucibles for melting gold, silver, and other metals. Of the various crucibles made those formed of plumbago have acquired the highest reputation, being used almost exclusively in the European mints and other government works. The plumbago used must be very pure—some of the samples average more than 98 per cent. carbon; but in every case it is subject to the most careful scrutiny by experienced workmen, who throw out any doubtful pieces. The plumbago then undergoes the process of grinding to different degrees of fineness, special regard being had to the metals for which the crucibles are intended. In like manner varying proportions of clay are mixed with the plumbago which are nicely adjusted to the prevention of the formation of a slag and of the oxidation of the plumbago. The compound mass is then worked like clay in the pottery manufacture. (Handbook to Exhibition f 1862, I. 32.)

If the quality of the plumbago found here should prove suitable for the treatment above described, and the quantity be sufficient, a valuable addition may be made to the resources of the provincial potters.

Sulphur. Specimens of a greyish black substance with a slight metallic lustre from an unknown locality, examined by Prof. Lawson, in 1866, proved to consist of native sulphur, which had not been previously known to occur in the province. In the first specimen examined the dark colour was found to be due to plumbago and not to the presence of metallic oxides or sulphides as is commonly the case in Sicilian specimens. The substance was very light, specific gravity 2, and when heated burned with the characters of sulphur ; it would readily afford pure sulphur by a simple process. In the same year Prof. H. Y. Hind mentioned to me that native sulphur had been found in very small amount at the Mount Uniacke gold mines : in one case a crystal was obtained but unfortunately lost. In November of the same year Mr. Nash showed specimens of " sulphur ore " from Cape Breton, at a meeting of the N. S. Institute.

Ores of Sulphur. Under this name mention is made in the British mining records of the common iron pyrites employed in the separation of sulphur and the making of sulphuric acid. This mineral contains 46 iron and 54 sulphur, in round numbers, per cent. It is a very common constituent of many rocks, is found in all the gold districts and appears to exist in large quantity in several localities abounding in slates : if it should prove to be really abundant in any favourable situation it would be well worth attention ; enquiry after deposits of it has been made of me on the part of a chemical manufacturer in New York who said that pyrites of 40 per cent. is worth there about $10 a ton. Some idea of the amount of the mineral raised and used in Britain will be got from the following details from the Mineral Statistics of the United Kingdom and the Statistics of the Alkali Trade.

Tons.
In 1861, the amount of pyrites raised in United Kingdom was... } 125,135, value £79,715.
In 1865, do............................114,195
In 1866, do............................134,402

In 1861 there were also imported 93,528 tons, and in 1864 the importation greatly exceeded the production. In 1862, 264,000 tons were used in the alkali trade. In 1862 Irish pyrites was worth 28 shillings sterling per ton. As before mentioned, pyrites poor in copper is first roasted for its sulphur and then smelted for copper. As regards its occurrence in this province, Mr. Poole states (Report, 1863) that on Long Island, Lunenburg county, the slates are very full of pyrites in cubes many of which are half an inch on the face; that the slates of New Germany hold abundance also, and that in the N. Queens district there is a ridge of quartz twelve feet wide, with pyrites: he also names many other localities in the western counties. Dr. Dawson mentions a bed of pyrites a foot thick at Montague and Dr. Gesner reports beds of it in the banks of Bear River, Digby county. Large and numerous crystals of it are found in the slates about Bedford Basin, **and a few miles from Windsor it may be** " got by the bushel " in such crystals as I have had given me by my informant. Mr. Barnes tells me he **knows a** bed of pyrites some 18 feet in thickness.

CHAPTER VI.

IRON ORES.—MINERAL PAINTS.

Iron Ores. The province is abundantly supplied with a variety of such ores as are employed in smelting: a few only of their localities have been the scenes of mining operations. These have been conducted almost exclusively in connection with smelting works, of which only one is now in activity; it produces a very fine quality of charcoal iron. Mr. Barnes, in reporting in 1866 on a deposit of iron ore in Colchester Co. of which further mention will be made, gives a summary of his observations on the iron ores of which it will be useful partly to quote the substance here as shewing generally the nature of the ores. "Having given my attention for some three years past to the economic value of the iron deposits of Nova Scotia and Cape Breton I have noticed the following varieties which probably embrace all at present **known to occur** in quantity.

"*Bog Ore.* Occurring principally in the coal districts of Sydney, Pictou, and Springhill—in such limited quantities as to be of little practical importance.

"*Clay Iron Ore.* In thin beds or bands, and nodules, chiefly in the lower coal formation: contains from 20 to 35 per cent. of metal. It occurs in Cape Breton in larger quantities than any other portion of the province, and of richer quality. Nowhere, however, **have I** seen it equal in abundance to similar deposits in the United Kingdom, where it is the chief ore smelted, and there it occurs in the same formations as observed here. Owing to the rich deposits of other ore I doubt much if this class will pay to work for many years, unless used as a mixture in the furnace with richer ore to produce different grades of pig iron.

"*Brown Hematites.* (*Limonite and Goethite.*) These ores are by far the most abundant and productive in the province, occurring in veins, beds, and masses at the base of the lower carboniferous strata.

"*Red Hematite and Specular Ore.* These are more widely spread through the province than any other class of ore I know of; they occur in rocks of greater age than the brown ores. Although I have examined these ores in many and widely separate localities I have never seen but a scanty and unimportant body of them." (A arge specimen of specular iron ore is shewn in the Provincial Exhibition from Guysboro' Co.; the ore is said to be abundant.)

"*Iron Sand. (Ilmenite)* occurs in several localities in Cape Breton, and also universally in the surface gravel along the Atlantic coast. It is not at present used but may hereafter be made available. Some varieties that I have seen are almost identical with the Taranaki sand of New Zealand largely imported into England."

Magnetic Iron Ore. This may be added to the ores existing in useful amount, as shewn hereafter.

Black-band Iron Stone. This may also perhaps be included as a useful ore as Mr. J. Campbell thinks he has found abundance of it in beds 3 or 4 feet thick; it is an impure carbonate of iron. I place here for reference an analysis of a specimen taken from an English work:—

ANALYSIS OF BLACK-BAND IRON STONE.

Protoxide of iron	53.03
Lime	3.30
Magnesia	1.77
Silica	1.40
Alumina	0.63
Peroxide of iron	0.23
Calcareous and bituminous matters	3.03
Water and loss	1.44
Carbonic acid	35.17
	100.00
Metallic Iron	41.24

Acadia Charcoal Iron Works. This is the name given to the only iron mining and smelting establishment in the province. The works are situated at Londonderry, Colchester county, and a detailed description of the nature and mode of occurrence of the ores is given in Dawson's Acadian Geology, from which I proceed to take some of the more immediately important statements; to these I add some account of the works kindly furnished in part to myself by Mr. Jones, the manager, after a visit I paid the establishment in 1861, and some analytical and other details of interest.

The iron ore worked is found in a vein of ferruginous magnesian limestone, a variety of ankerite, which extends along the south slope of the Cobequid Hills, and which has been most carefully explored in the vicinity of Folly and Great Village Rivers. At the site of the Acadia Mine furnace, in the western bank of the Great Village River at the junction of the carboniferous and metamorphic rocks, a thick series of grey and brown sandstones and shales, dipping to the south at an angle of 65° and 70°, W. meet black and olive slates, nearly vertical and with a strike N. 55° E. The vein is well seen in the bed of the stream, and also in excavations in the western bank which rises abruptly 327 feet above the river bed. In the stream-bottom it presents the appearance of a complicated net work of fissures penetrating quartzite and slate and filled with ankerite, with which is a smaller quantity of red ochrey iron ore and of micaceous specular ore. In ascending the western bank of the stream the vein appears to increase in width and in the quantity of the ores of iron. In one place a trench shewed a breadth of 120 feet. In some parts of the vein, the ankerite is intimately mixed with crystals and veinlets of yellowish spathose iron. The red ochrey iron ore occurs in minor veins and irregular masses dispersed in the ankerite. Some of these veins are two yards thick and the shapeless masses are often of much larger dimensions. Specular iron also occurs in small irregular veins, and in disseminated crystals and nests. At one part of the bank there appears to be a considerable mass of magnetic iron ore mixed with specular ore. The general course of the vein at the mine and further to the east is S. 98° W., the variation being 21° west. At the mine the course deviates about 33° from that of the containing rock, elsewhere the deviation is less, and there is an approach to parallelism between the course of the vein and that of the rock formation of the hills as well as that of the junction of the carboniferous and metamorphic systems. The vein for a space of 7 miles along the hills is always found at the distances of from 300 yards to one-third of a mile northward of the last carboniferous beds and always in the same band of slate and quartzite.

Westward of the Acadia Mine the course of the vein is marked by the colour of the soil for about a mile, as far as Cook's Brook where the outcrop of the ore is not exposed but large fragments of specular iron have been found, and a shaft sunk on the course of the vein has penetrated more than forty feet of yellow ochre containing a few rounded masses and irregular layers of ankerite. Specimens of specular ore and ankerite have been received from the continuation of the same metamorphic district as far west as Five Islands, twenty miles distant from the Acadia mine.

On the east side of the west branch of the Great Village River, the vein is not so well exposed, but indications of it can be seen on the surface as far as the east branch of the river, in the bed of which it has not, however, been found to continue. Further eastward, on the high ground between the Great Village and Folly Rivers, indications of the ores of iron have been observed, especially near the latter, where in two places small excavations have exposed specular and red ores, and where numerous fragments of brown hematite are found on the surface. On the elevated ground east of the Folly river, the vein is again largely developed; at one point 10 feet of red iron ore were seen without exposing the north side of the vein. On the surface in this vicinity are large fragments of brown hematite which mark the course of the vein. In a second excavation the red ore was more largely mixed with micaceous and includes large rounded blocks of ankerite and angular rock-fragments. The width here exposed was 15 feet and neither wall was seen. Still further east, on the property of C. D. Archibald, Esq., on equally elevated ground, three excavations have shewn a still greater development of the vein, one trench 53 feet long, nearly at right angles to the course of the vein, shews in its whole length a mixture of red and specular ores with ankerite. In the bed of the Mill Brook about 2 miles east of Folly River the vein attains a great thickness in the eastern bank, here it consists of a net-work of fissures filled with ankerite: it was found in the bank of another brook still further to the east, and though not traced further there was no doubt entertained of its continuance to a great distance in that direction. The deposit was pronounced to be wedge-shaped, the largest and richest portions occurring on the surface of the highest ridges containing an immense quantity of valuable ores of iron, and from its irregular character opposing difficulties to the miner. From the elaborate report of J. L. Hayes, of Massachusetts, it appeared that even if the cost of the ore amounted to 4 dollars a ton, there would be an advantage over the average of American localities; at some establishments smelting at a profit the ore cost from 5 up to 10 dollars a ton, the average in Pennsylvania being about 7 dollars. The cost of ores at some of the Swedish and Russian furnaces is still greater. In certain parts of the Urals the **ores are carried by** land from 40 to 80 miles. In Sweden some forges are supplied with ores transported, by land, the lakes **and the sea,** distances exceeding 370 miles. I visited the mines in **1861, with R. G.** Haliburton, Esq., secretary to **the N. S.** Commissioners for the London Exhibition of 1862, and saw with great interest the admirable arrangements by which a large amount of work was being done, and having requested E. A. Jones, Esq., the manager, to favour me with some descriptive details, I received the following valuable account of the history, progress, and nature of the establishment.

"The Acadia Iron Works were commenced in 1849, and the first **iron** was made by the Catalan forge in 1850. In 1852–3 a blast furnace was erected for the manufacture of pig iron, the Catalan forge being abandoned. Up to the time of my arriving in the province, in the summer of 1857, there had been **made** altogether about 1000 tons of **iron,** from about 4000 tons of ore. Since that time to the present (1861) we have made about 4000 tons of iron, using about 9000 tons of ore. Our present make of bar iron is at the rate of 1200 tons, of an economic value of about £24,000 per annum. The ores we use are a hematite, yielding about 48, and a **brown and red** oxide yielding **about 40 per cent of** iron. The ores are somewhat **refractory:** this arises mainly **from the** presence of a stone **mechanically** mixed through the ore **which is very** difficult to act upon in the blast furnace. It requires about 160 bushels, imperial, of charcoal, and 200 bushels of limestone (this is found in the vicinity) used as a flux, to smelt one ton of pig iron, and about $8\frac{1}{2}$ cords of wood to convert the pig into bars. The wood used is required to be perfectly dry: for drying it we use artificial means, and also house a large quantity in sheds for winter use—as much, this year, as 1000 cords. We have one blast furnace and 3 puddling furnaces, with one reheating furnace; the pressure of blast used is about 4 oz. to the square inch, and the quantity of air about 2000 cubic feet in a minute. We now employ about 230 men; our expenditure for wages, etc., at the works will average about £1200 a month. The iron made compares very favourably with the best metal brought to market from any part of the world for the same purpose, namely, the manufacture of steel. Thus, the Swedish iron, of which there are many varieties, brings from £12 to £25 sterling, one brand as much as £30 to £34, in the Sheffield market. The Acadia iron is worth about £16 sterling per ton, so that it compares with the average Swedish. It should be added that at Londonderry a less expensive mode of manufacture is adopted than that usually pursued in Sweden, on account of the high price of labour in this country." Writing in January 1864 Mr. Jones further informed me: "I have little to add to the account I gave you of our works, and excepting in the increased production there is no change; we shipped in 1863 900 tons of bars and 402 tons of pig iron, of the aggregate value of $85,000. We are now driving an adit into the mountain at the head of the river, which will enable us to ascertain the nature of the deposit of ore at about 100 yards depth. The value of our iron as compared with English of the best quality, is but estimated from the selling price:

	£	s.	d.			
English Pig Iron (Staffordshire) average	4	0	0	stg.	per	ton.
Acadian ,, ,, ,,	7	0	0	,,	,,	,,
English Bar Iron (Staffordshire) ,,	9	0	0	,,	,,	,,
Acadian ,, ,, ,,	15	10	0	,,	,,	,,

As compared with Swedish iron, our bars rank with the best qualities, there being *but one* iron which is considered superior for *steel*; our bars are all used for this purpose and the demand is steadily increasing. The pig iron is used principally for making railway wheel tires, for which purpose it is well suited, being, when converted into malleable iron, very compact and not liable to wear by attrition. This year we have for the first time shipped to the United States, where the bars are coming into demand for making steel. Dr. Percy has found titanium in our iron in considerable quantity. I may add that we have purchased a neighbouring property on the same mineral range, with a view to extending our manufacture."

In 1865 there were 250 men and boys and 25 horses employed on a daily average. The following statement shews the make of pig and bar iron for several years:—

	Pig Iron Made.	Bar Iron Made.
1862	1150 tons.	945 tons
1863	1251 ,,	911 ,,
1864	1663 ,,	1198 ,,
1865	1784 ,,	1633 ,,
1866	2124 ,,	1093 ,,
1867	2068 ,,	421 ,,

The falling off in bars for the last two years is due to the depressed state of the iron trade in England.

In all there have been made at the Acadia Works 15000 tons of pig and 7000 tons of bar iron, of the aggregate value of $1,000,000 (one million dollars).

The following analytical details will be found of great interest.

Analysis of Pig Iron made at the Acadia Mines published in Percy's Metallurgy:—

	I.	II.
Carbon	3.50	3.27
Silicon	0.84	0.67
Sulphur	0.02	0.01
Phosphorus	0.19	0.28
Manganese	0.44	0.37
Iron	94.85	95.70
	99.84	100.30

The manganese contained a sensible amount of cobalt. No. 1 was a coarse grained, No. 2 a fine grained, metal. Mr. Jones understands that Dr. Percy said titanium had been found in his laboratory but he has no figures or definite information on this point.

Analyses of ores from Acadia Mines made in 1865 by Messrs. Woodhouse and Jeffcocke, of Derby; No. 1 is from Martin's Brook, where the ore now worked has been chiefly extracted; No. 2 from a different part of the property;

ACADIA CHARCOAL IRON WORKS.

	I.	II.
Peroxide of iron	80.36	84.13
Carbonate of lime	4.80	4.70
Carbonate of magnesia	2.26	1.80
Silica	4.60	5.20
Alumina	5.13	
Water	2.36	3.20
	99.51	Phosporic acid none
		99.03
Metallic Iron	56.25	58.90

The material now used as a flux is one which I analysed for the company last autumn; I found in it:—

Carbonate of lime	51.61
Carbonate of magnesia	28.69
Protocarbonate of iron, and some peroxide of iron	19.57
Gangue and silica	0.13
	100.00

Of this Mr. Jones says there is an extensive deposit in the neighbourhood and that it is very valuable as a mixture in fluxing clay or silicious ores.

"The veins of ore, which are now (1868) pretty well defined, are two in number, and continue as we go deeper with our workings much *the same as they were found on the surface*. We have now one adit 250 yards in the ore at a depth of 40 yards below the surface, and another 160 yards across the measures towards the ore at a depth of 80 yards. I do not know that I can add anything more except perhaps to say that the Intercolonial Railway will pass through the iron ore district and connect it with the Springhill coal field, a distance of 24 miles, which will assist very much in developing the iron interest."

The latest particulars with regard to the ores, some of which will explain the passage in Mr. Jones's letter which I have put in italics, are given in a paper by Rev. Dr. Honeyman on the geology of the mines in the Transactions N. S. Institute for 1867. They are to the effect that the ore worked till recently was specular ore derived from a bed about 3 feet thick, and hematite from a bed of variable thickness and unknown depth. The brown hematite is now the only ore available, the specular having been apparently exhausted, while another great bed of hematite has been found nearly as large as the bed already referred to. The two beds are now known as the north and south, their strike is E. and W., their dip 80° S. At Martin's Brook they appear about 30 feet apart. The maximum thickness of each bed is 20 feet, the average of the north is 5 feet, of the south 4 feet. Very often the beds are interrupted and disappear; they have been traced for 12 miles. It was supposed that the hematite was an altered ankerite and that it would only be found in the top of the vein. It is now certain that the hematite at 100 feet depth is of precisely the same character as in excavations near the surface. Cavities with botryoidal crystalizations were found in the roof of the level as well as in excavations above.

Mr. A. Ross informs me that about 4 miles east of the Acadia Mines about 500 tons of the best iron ore have been raised from a pit and shaft sunk to a depth of 46 feet.

As mentioned above the iron made at Londonderry is especially adapted to the manufacture of steel; for this application the company has an establishment at Sheffield, in England. Articles made at these latter works have been shewn in successive Great Exhibitions, from that of 1851 to that of Paris 1867, along with the ores found and iron produced in Nova Scotia. In 1851 the company received a Gold Medal in London for their Acadian cutlery. In 1854 a prize was obtained at the Local Exhibition held in Halifax, for ores and iron. In 1862, at the International Exhibition of London, Dr. Honeyman reported; "The excellent specimens of iron sent by Mr. Jones did not receive at the hands of the jurors the consideration they appeared to deserve, if we are to be guided by the opinion of those who professed to be judges of their quality. If the pig iron bars and ores sent by Mr. Jones had been accompanied by a representation of the character, quality, and application of the Londonderry iron, I have not the least doubt that the united representation would have received the jurors' award. I may state in this connection that the *Times Correspondent* took occasion when writing on these ores to make a rude attack on the Board of Provincial Commissioners for having sent to our court the specimens of our ores of iron. I replied, but the journal did not condescend to insert my reply. The correspondent of the *Morning Star* in an excellent article on our court, took up the question and severely reproached the ignorance of the *Times Correspondent*. R. G. Haliburton, Esq., Sec. to the Commissioners, reported: "It is to be regretted that one of the Directors of the Acadia Charcoal Iron Co. was elected juror on iron, as the specimens of its cutlery as well as of the ores employed were excluded from the competition. It is satisfactory to know, however, that a medal would have been awarded but for the circumstance referred to. The acting commissioner in England, A. M. Uniacke, Esq., on seeing the article in the *Times* respecting the iron shewn by us, wrote to that paper to explain that none of the ore to which it objected was to be seen in the Nova Scotian court. The specimens that were decried in no very measured terms were in reality the best in our department and realize in the English market a price second only to the very best Swedish brands. Mr. Uniacke in an official letter asked the *Times* to correct the mistake, but his communication was not honored with an insertion nor was its receipt acknowledged. It might naturally have been expected that a request so reasonable would have been readily granted, not as a favour to the colony, but as a concession to truth." The ore objected to was some from a distant part of the province, viz: Nictau, which as will be mentioned hereafter, contains phosphorus in greater quantity than is desirable for making the best iron. At the Dublin International Exhibition of 1865, Dr. Honeyman reported: "E. A. Jones, Esq., manager of the Acadian Iron Mines, is awarded a medal for Pig Iron and Hematite. It certainly adds to the value of this metal that the decision of the jury was facilitated through the kindness of Lady M'Donnell in permitting us to exhibit a beautiful case of cutlery made of Acadian steel, presented to her by Mr. Livesey. It was a fortunate circumstance that this case was exhibited as we did not receive the large case of cutlery from Sheffield which we had expected. There was another fortunate circumstance connected with this article of exhibition. One of the jurors adduced the objection and misrepresentation in regard to the manufacture and quality of the Acadian iron which had been offensively set forth by the *Times*, in 1862. In this case, however, I succeeded satisfactorily in meeting the objection and the medal was unanimously awarded and the Acadian iron restored to its proper position." As for the Paris Exhibition, of 1867, it appears from the catalogue that the representation, though small, was instructive. The ore, brown hematite, was shewn in a great variety of interesting and curious forms. The quality of the metal was seen in pig iron, the malleable being represented by a bar. In this state it is found to be as well adapted for shoeing sleds and sleighs as a great proportion of the steel imported to the province. Specimens of cast and puddled steel, also exhibited, manifested superior density and tenacity. These were a part of the first results on attempting to make steel. The axe and chisel shewn were also made at the works from iron and steel represented. It is not surprising to find Dr. Honeyman, Secretary to the Provincial Commissioners, reporting. "I am rather disappointed at not finding the admirable illustration of the Acadian Iron Works in the list of awards, especially as I have heard these contributions much commended by the jurors when they were in process of examination."

Hematite of Brookfield, Colchester Co. A deposit of iron ore in some respects apparently similar to that at Londonderry is found in another part of the same county. This I shall describe chiefly from the report of Mr. Barnes, some portion of my own report made after prospecting the locality, and the results of my analysis of the ore, being added. Mr. Barnes says: "Oct. 20th 1866; I am happy to afford any information I can from the results of my investigations of a tract of iron land, controlled by Charles Annand, Esq., at Brookfield, in the township of Truro and county of Colchester. The land is on the "Nelson farm" which consists of three blocks of land and contains on the whole about 400 acres; it is situated about 2½ miles from the Brookfield Station on the Halifax and Truro Railway, and ten miles from the town of Truro. The strata of the surrounding locality are lower carboniferous limestones, shales and coarse conglomerates, resting on red, yellow, and dark sandstones of a similar character to those of Londonderry in the same county, and of Springville, Pictou county, where brown hematites also occur. Masses of ore of remarkably rich quality lie scattered over part of the land in great profusion; they vary from pieces as small as a marble to blocks weighing 3 and 4 tons. Detached fragments in equal profusion mark the underlying deposits in Londonderry and Springville.

The openings at present made shew the bed rock to be the matrix of the scattered masses of ore. In one, the principal cutting, a course of red and yellow sandstone has been penetrated containing rich ore in situ of from two to three feet wide. It is at this point that further explorations should start as from the direction of strike of the beds, and the position of the scattered masses, a large and rich deposit of ore is close at hand. Indeed, from the abundance of ore on the surface, and the little abrasion it appears to have suffered, I consider that the indications of an extensive deposit are greater than even at Londonderry Mines. The mine is on lands granted free from all control of the Crown and therefore no interference can be exercised by the Department of Mines. All the masses of ore I examined are uniformly pure limonite containing from 60 to 65 per cent. of metallic iron. This quality of ore, from its freedom from impurities, is eminently adapted to the manufacture of the finest varieties of wrought iron and steel, for the production of which charcoal in great quantity can be obtained from the heavily timbered lands extending eastward for 30 miles to the Middle River of Pictou, and southwardly for 18 miles. The ordinary price paid at the works at Londonderry is 7 cents per bushel for hard sound charcoal from the maple, birch, and beech, delivered at the works, and sometimes carted 10 or 12 miles. This charcoal, therefore, weighing $22\frac{1}{2}$ lb. to the bushel, would cost nearly $6.30 per ton of 2000 lb. The average amount of charcoal used to produce one ton of iron varies very much with the nature of the ore and the shape and size of the smelting furnace. With the ore in question 160 or 170 bushels will be the amount required, costing therefore $11.20 to $17.90 per ton of iron. In working a furnace making 8 to 10 tons of pig per 24 hours, 8 men are required. The value of the charcoal iron, according to the latest quotations is $35.33½ per ton in the London markets. The pig iron produced at the Londonderry Mine is sold at $30 to $36, while their bar iron is as high as $50 to $60 per ton. The situation of the property is such that works could be established on the ground for the purpose of smelting or the ore could be taken by a tramway $2\frac{3}{4}$ miles along a level line to the Brookfield Station and thence on the railroad to the neighbourhood of the Albion Mines, a distance of 30 miles, where coal and coke could be procured in abundance and coke iron made at less expense than at Brookfield. The charcoal would cost a trifle more if delivered at the **Albion** Mines but as, when the mine is worked, it will be required probably to make more coke than charcoal iron a smelting work at or near the East river of Pictou would be most advisable. Besides the abundance of fuel, an extra advantage would accrue from the facility of obtaining other classes of ore for mixing, and easier and cheaper outlet for the iron made, not only to various parts of this province but to the other North American provinces. The importation of Scotch and English charcoal iron into this province is annually about 3700 tons and the cost, freight and duty (5 per cent,) paid, $19 to $23 per ton. I have carefully gone over the expenses incidental to the manufacture of both charcoal and coke pigs and believe that both varieties could be made from the ore of Brookfield and sold at $6 or $7 under present import prices, after paying a dividend of 20 per cent. upon any capital that might be required in the erection of furnaces and developing the mine."

The importation of iron of various kinds is very rapidly increasing: much of course is imported for railway purposes and, in addition to the large foundries and machine shops in Halifax, there are several foundries throughout the province at which much iron is consumed, there are also four nail factories in Halifax and Dartmouth.

Amount of Iron and Steel imported. The following statement shews the value of iron and steel imported into the province. It is taken from the Trade Returns where the amount is given sometimes in tons, sometimes in pieces and packages, so that I cannot give the quantity :—

Value of Iron and Steel imported into Nova Scotia, in the years ending 30th Sept., under the head of Hardware.

	1864	1865	1866
Class I. Cutlery, Sheffield Ware, Stoves, etc................	$301,245	664,347	672,514
Class II. Iron and Steel in Bars, Wire, and Machinery for Steamboats, Mills, etc.	$297,864	473,497	616,496
Class III. Pig Iron, Railway and Scrap Iron, etc	$38,951	271,645	321,409
Total..................	$638,060	1,409,489	1,610,419

To the foregoing report on the Brookfield deposit I may add from my own the following extracts: "The great appearance of the ore is in the form of masses and boulders lying in great profusion in about an east and west line on the south side of the hill. About 10 chains east from Nelson's house an opening has been made in hard sandstone which is evidently the bed rock in position, and has an east and west course and almost vertical dip to the north or into the hill. This rock contains a good deal of iron ore in strings and irregular veins and stalactitic masses in cavities. One side of the opening showed the rock for about 5 feet in thickness. Judging from the appearances here the rock should contain about 40 or 50 per cent. of ore. The character of some of the ore was precisely similar to that of the loose masses which extended for about 6 chains. From what I saw there must be very many tons lying on the surface. The masses are of various sizes, two in particular being of very large dimensions; one of them measured 3 feet in length, and 2 feet 3 inches in breadth and the same in depth. I was assured that the summer growth of weeds and brush prevented a great proportion of the surface samples being seen. The ore is evidently very rich, many of the masses, probably the majority, appearing to consist of nearly pure limonite, which contains about 60 per cent. iron. The poorest specimens, some of which are mixed with barytes, would yield a large per centage of iron. The geological position and appearance of the ore are very similar to those of the Londonderry Mine in the same county. It appears to me there must be a large body of ore in the hill. The position of the ore is favourable for mining purposes, being in the slope of a hill, and should it be deemed advisable to smelt the ore on the spot, any desired quantity of charcoal and limestone can be readily obtained in the vicinity, and sandstones in abundance are at hand for building purposes. I carefully examined a sample of the fibrous variety of ore taken from a large piece, and found it remarkably pure. The amount of phosphoric acid present was a mere trace; there was a doubtful trace of sulphuric acid; lime, magnesia, except in exceedingly small traces, and manganese were absent. The quantity of alumina was so small that I did not consider it worth separation; it probably did not amount to one half a per cent. The result of analysis was:—

```
Water ..................................11.36
Silica and gangue........................ 1.54
Phosphoric acid..........................trace.
Magnesia ................................trace.
Peroxide of iron with a very little alumina.87.10
                                        ──────
                                        100.00
```

The oxide of iron obtained is equal in round numbers, allowing one-half a per cent. for alumina, to 60 per cent. metallic iron.

Although manganese was not found in the particular sample analysed it was proved to exist in small quantities in another specimen of the ore; the presence of this metal, in certain quantity, is rather advantageous in making steel, an application for which this ore is especially adapted."

Ores of East River, Pictou Co. Various iron ores are found in the vicinity of East River, Pictou. From some of these iron was formerly made to a small extent at Webster's, about five miles above the mills on M'Lellan's Brook, from ore obtained at M'Donald's. Last summer I had given me at the Albion Mines a piece of iron from a mass which was smelted in a small furnace there thirty years ago: a portion of the same had been used in making the stampers of a quartz mill at Waverley and had been pronounced to be " ten times more durable than Belgian iron."

Dr. Dawson describes the ores of this locality to the following effect :—" From the abundance of boulders of brown hematite scattered over the surface of the lower carboniferous rocks in the East River it would appear that veins of that rich ore exist in these rocks. The outcrop of these veins has not yet been observed, and as the country is much covered by drift materials it may prove somewhat difficult to discover them. The presence of these ores in connection with a large bed of peroxide of iron in the older slates leaves little doubt that were other circumstances favourable iron works might be established on the East River without any fear of deficiency in the raw material. The bed of ore in the older slates at East River appears to be of great magnitude. Though the ores are less rich than those of the Cobequid Mountains, being siliceous, the deposits are likely to be more continuous and persistent. The great bed of ore on the East River of Pictou is especially worthy the attention of capitalists as it is only ten miles distant from the

Albion Coal Mines and is in the vicinity of abundance of limestone and building stone. The hematite and clay iron ores found in the carboniferous rocks might be used with the ore of the great conformable bed in the slates which consists of specular iron firmly cemented together and intermixed with siliceous and calcareous matter." This district was examined by Dr. Honeyman, in a Geological Survey commenced for the Provincial Government in 1864, and a few details from his report will be useful. "Dr. Honeyman devoted the greater part of the season favourable for field work, to the east branch of East River, Pictou Co., and the Lochaber district, Antigonish Co. These regions are peculiarly interesting, as they present unmistakeable indications of the existence of metallic deposits of economic value. These have long been the objects of anxious search. The General Mining Association and others have spent much time and money in attempts to win the more important metallic veins. With a view of ascertaining the exact position of the veins attention was turned first of all to the great vein of brown hematite at the east branch of East River. Indications of its existence were found through a course of 5 miles where rights of search had been secured by Nova Scotian and American companies. The geological position of this vein is upper silurian, being different from the veins of peroxide of iron and sulphide of copper of the Lochaber district, which are devonian, and apparently so from the vein of similar ore on the southern skirt of the Cobequid." In connexion with this survey several minerals were submitted to me for analysis, a report on which was published in the Journals of the House of Assembly for 1866. One specimen was brown hematite from East River of Pictou which I found to be very pure ore, it gave me the following results:—

Peroxide of iron, with trace of phosphoric acid....84.54
Alumina and phosphoric acid 0.19
Sesquioxide of manganese........................0.76
Magnesia 0.43
Water..11.41
Siliceous gangue 2.22
Carbonic acid and loss.......................... 0.45

                                            100.00

Metallic iron per cent..........................59.17

The ore is obviously rich and valuable, it contains within one per cent. less than the amount of iron that, as mentioned in speaking of the Brookfield ore, characterizes the class to which it belongs.

Brown Iron Ore from Lochaber. A specimen of ore was at the same time sent from the district just spoken of in Dr. Honeyman's report; it gave me :—

```
Peroxide of iron and traces of alumina............68.45
Water .............................................11.12
Sesquioxide of manganese.......................... 4.73
Phosphoric acid................................... 0.33
Lime.............................................. 0.34
Magnesia ......................................... 0.32
Siliceous gangue .................................13.86
Carbonic acid and loss ........................... 0.85
                                                 ------
                                                 100.00
Metallic iron, nearly.............................48.00
```

This is also a valuable ore, the amount of manganese present is advantageous in neutralizing the phosphorus which however is not unfrequently found in ores of this class in much larger quantity than shewn in the analysis given. Of the iron ore here Dr. Honeyman reported that in the sedimentary rocks he found carbonate of iron, and in a series of brownish red strata of great thickness and width, of devonian age, micaceous iron ore was widely but thinly distributed and the blue slates of the same period at Polson's Lake and South River Lake contained numerous veins of carbonate of iron (ankerite?) with sulphide of copper and veins of oxide of iron, the former being undoubtedly a continuation of the veins which produced the masses of cupriferous oxide of iron which have long attracted the attention of geologists. R. G. Haliburton, Esq., has lately discovered and tested a workable deposit of very rich specular iron ore on the line of Railway near East River, Pictou county.

Ores of Annapolis County. At Moose River and Nictau River are beds of ore, which Dr. Dawson describes as of the same nature as that in the slate-bed of East River of Pictou, consisting of conformable beds in the lower devonian slates. At Clementsport, Moose River, about six miles south-west of the town of Annapolis, the iron ore is found in a magnetic condition and holding fossil shells; the bed is nine feet wide. Smelting operations were formerly carried on here and just before the commencement of the late American war they were resumed after a stoppage of thirty-three years; in 1862 five tons of iron a day were being turned out. (Knight's Prize Essay on Resources of Nova Scotia.) In a year or so the works were again closed on account it is said of the death of the senior partner and the affairs being influenced by the war; another reason assigned is that a lawsuit is pending. The hot blast system was used in a cupola furnace: it was said the largest wheel in the province, one of 75 feet in circumference, was in this establishment.

At Nictau River, some 30 miles east of Clementsport, the ore is also fossiliferous: it consists of specular iron which has been in part rendered magnetic; specimens which I have examined shew distinct polarity. In this ore (and no doubt the preceding) we have a case where the magnetism depends on the state of aggregation and not on the chemical composition of the ore (Nicol's Mineralogy, p. 398). This is a point of importance because true magnetic iron ore is essentially different from specular ore, and frequently yields iron of a superior character; it contains about $2\frac{1}{2}$ per cent. more iron also, pure specimens being compared, the rest of the mineral in such cases being oxygen. I visited the locality some years ago when there was a great deal of iron being made. I am indebted to the Rev. Dr. Robertson, whose rectory is a few miles from the mines, for some valuable information respecting the minerals of this part of the province; on this subject he gave some interesting details of which the following is essentially the substance: —" The Nictau mines have been worked for many years, and extensive works have at great expense been erected for smelting the ore, but at present they are in a state of inaction. Some conjecture that the difficulty and expense of carrying the products of the furnaces to the landing place, 11 miles off, may have been the principal reason of the present inactivity. The vein that has hitherto been worked is situated on the east side of the River Nictau, and is intermixed to a large extent with petrified marine shells. These shells still contain their natural calcareous properties. They are very clearly marked, leaving a well defined impression in the matrix of the minutest lines. They are often found in clusters so compact and homogeneous that one might imagine the whole to be formed originally from one **vast bed of** shells. The vein is about ten feet wide or thick **and is** found to extend for some three or four miles. I once saw **an** analysis of the ore but I forget the exact proportions. There **was** however a small per centage of phosphorus detected and this fact is supposed to depress the marketable value. An engineer from the United States was here looking at the mines and works some time ago and he said that a corrective might easily be found for that." With reference to the phosphorus I have learned from another source that **the ore** contains phosphorus, and, as mentioned before, the quantity is said to be injuriously large. It is in all probability one cause at least of the inferiority of the Nictau iron. The corrective mentioned above is no doubt manganese ore which as hereafter shewn is proved by Dr. Calvert to neutralize the injurious effects of phosphorus and of silicon, five or six per cent. of manganese making good mercantile pig iron even in the presence of one or two per cent. of phosphorus. The ores might be taken by rail, as soon as the Windsor and Annapolis line is opened, from near Wolfville or Kentville, Kings county, if those there met with should be found sufficiently abundant, or even from Hants, and the same results might be obtained as by the Cleveland iron smelters in England who overcome the cold shortness of their iron, due to phosphorus, by the addition of manganese ore. The Nictau iron ore has been examined chemically by Dr. Dawson who describes it as being siliceous and containing 55.3 per cent. of iron. In 1858 the quantity of iron exported was 744 tons, value $2,375, and in 1859, 1125 tons, of the value of $14,790, (fourteen thousand seven hundred and ninety dollars.)

Magnetic Iron Ore. I was taken several years ago to see the outcrop of a deposit of this ore which Dr. Robertson now tells me is about 8 feet thick and runs north-east and south-west across the line of the mountain range. The locality at which I saw it was on a hill on the west side of Nictau Falls. Dr. Robertson is of opinion that the ore is far superior to the shell ore before mentioned. It appears to him to be very compact and to contain a very large per centage of pure iron. It has never been analysed. I did not obtain specimens when I was there but I have received some from the county, and perhaps the locality referred to, which are of excellent promise.

Iron Ore of King's County. Magnetic iron ore is also said to be abundant in this county. I have received several fine specimens reported to be from near Blomidon and other parts of Cornwallis where trap rocks are found forming the North Mountain One of these specimens weighed several pounds and was sent to the Paris Exhibition. The ore was, according to Dr. Gesner, formerly sent to the United States; it was taken from a vein six inches wide, in the trap of Blomidon.

Iron Ores of Digby County. Dr. Robertson tells me there are deposits of specular iron ore near the Sea-wall, which is an embankment on the north side of St. Mary's Bay, about ten miles west of Digby. The same ore occurs in veins about two miles to the east and is again seen at Sandy Cove, fourteen miles to the west. Of specular ore at this last place Dr. Dawson says the quantity is not sufficient for mining purposes and that it occurs in brilliant little crystalline plates in a quartzose matrix projecting from the sides of cavities in fissures of the trap. Numerous veins of magnetic ore are met with at various localities in the trap of Digby Neck.

Bog Iron Ore from Antigonish County. Among the specimens I examined in connection with Dr. Honeyman's Geological Survey was one which indicates a valuable deposit if such ore is found in quantity. It contained nearly 65 per cent. peroxide of iron, equal to about 45 per metallic iron, with 18.30 water, about 7 of clay, and 5 per cent organic matter and a decided, but not unusually large amount of phosphoric acid.

Hydrated Red Iron Ore. Turgite. In specimens of iron ore brought to me, I think probably from various parts of Hants county, I have observed a red ore which is different from red hematite inasmuch as though equally red it contains about 5 per cent. of water which is entirely absent in true red hematite It is of importance to be aware of the existence of this ore since the presence of it in brown hematite, with which it is often associated, will add to the percentage of iron while it contains less iron than red hematite and the results of analysis may not accord with the appearance of the ore.

Titaniferous Iron Ore. As before mentioned, this ore, in the form of sand, is found in several parts of the province. It occurs also at Sable Island. I found titanium in the magnetic grains and Dr. Percy says the ferruginous portion, (the rest is quartz sand), is chiefly magnetic iron with a little titanium and a trace of chromium. A sample from Digby Co., procured by Mr. R. G. Haliburton, who understood the quantity to be large, I found to consist of grains of quartz sand and a mixture of magnetic and non-magnetic iron ore in the following approximate proportions :

Magnetic iron sand or Iserine	30
Non-magnetic iron sand or Ilmenite	56
Siliceous sand	14
	100

I found titanium in both forms of metallic sand and contented myself with proving its presence in large quantity in the whole without attempting the separation of it from the iron and magnesia. The analysis of these ores given in Dana's Mineralogy shew a very great difference in their respective richness in titanium : thus they contain

	Iserine.	Ilmenite.
Oxide of iron	91	91.5 to 46.4
Oxide of titanium	9	8.5 to 53.6
	100	100.0 100.0

The amount of the last named ore found in my analysis of the Digby specimen, it will be observed, is about 56 per cent.

Titaniferous Iron Ore of Sable River, Shelburne Co.: is reported to be found in a vein on the Atlantic Coast.

I obtained a specimen from Mr. R. G. Fraser reported from Musquodoboit as titaniferous ore. It consisted of a micaceous schist thickly impregnated with small crystals of magnetic iron in which I proved the existence of titanium in considerable quantity. I conclude that the ore is in part ilmenite.

The value of titaniferous iron ore for steel making has been of late years much insisted on in England though there are those who refuse to allow that it has any good effect. It was mentioned in Mr. Jones's last letter that Dr. Percy had found a good deal of titanium in the Acadia iron; this fact was brought out by a question of mine, my object being to ascertain how far this excellent steel-making iron agreed in this respect with Swedish and other irons noted for admitting of the same application. With regard to some of these Mr. Mushet states that if any chemist will be at the pains of analyzing the steel irons used in Sheffield, he will find that their market value is in exact proportion to their percentage of titanium; also that the Dannemora magnetic iron ore contains a larger amount of titanic acid than any other ores giving inferior brands of Swedish iron; that the celebrated Damascus blades are made from a highly titaniferous ore; that the Wootz ore of India is more titaniferous than that of Dannemora; that irons alloyed with titanium possesses a degree of body and durability unknown in ordinary good bar, and, finally, that first rate steel can only be made from iron containing titanium.

The durability of Acadia iron railway wheels may be recalled in this connection. Mr. Mushet even says one half a per cent of titanium may possibly constitute the excellence of steel, and that as all magnetic iron ores contain titanium the most impure ores of this class yield superior iron. This should be of interest to the owners of the magnetic iron ores in Annapolis and Kings counties before mentioned.

Mr. Struson, also an English ironmaster, entertained much the same views as Mr. Mushet; he said the remedy for certain difficulties in the working of titaniferous iron ores was to add more limestone and other fluxes, and that iron of various qualities could be produced at will, a soft malleable iron resulting from one assay, and from another, a fine grained silvery steel which when made into a chisel could cut any other steel in his possession. These experiments were made on titaniferous iron sand from Taranaki, in New Zealand, which is found in enormous quantities and no doubt closely resembles the iron sands of this province: by analysis this sand appears to consist of

        Protoxide of iron........................88.45
        Oxide of titanium and silica..............11.43
                                      99.88

The numerous difficulties attendant on the smelting of this ore have hitherto prevented its being employed in the making of iron.

It is, however, stated that good pig iron is now made from it by Mr. Martin, of London, by smelting it in small furnaces with coke for fuel. The examples of iron and steel made from it which have been exhibited are of a very high character, which is supposed to be mainly due to the presence of titanium. (Quarterly Journal of Science, Jany., 1866). Mr. Hodges has also recently smelted iron sand by moulding a mixture of it with pulped peat into bricks and heating these in a proper furnace. By this means malleable iron is readily obtained by a single operation. The remarkable process of Mr. Ellershausen for which patent rights have been or are being secured for all the leading countries in which iron is manufactured was at first understood to have exclusive reference to the working of the titaniferous iron sand so abundant on the coast of Labrador.

Mineral Paints. The name of mineral paints is given to the ochres and umbers consisting of peroxide of iron and manganese, existing in the hydrated state often mixed with sand clay and other minerals; and also less frequently to marls and clays containing peroxide of iron. These mixtures have been largely used in the province and have been exported from time to time in considerable quantity, occasionally in a manufactured state, from widely distant localities. The natural production of the soft powdery umbers and ochres can in some places be studied in every stage of alteration of the hard rocks from which they originate.

Ochrey Iron Ores of Londonderry, Colchester Co. These are found of various colours and in great abundance in the vicinity of the Acadia Iron Works. The original material appears in most cases to be the ankerite forming the vein stone of the iron ores. The ankerite is a hard sparlike mineral either white, yellow or brown: an analysis of each variety has given these results:—

	J. W. Dawson. White.	C. T. Jackson. Yellow.	C. T. Jackson. Brown.	H. How. Brown.
Carbonate of lime	54.0	43.80	49.20	51.61
Carbonate of iron	23.2	23.45 ⎱	20.30	19.59
Carbonate of manganese	....	0.80 ⎰		....
Carbonate of magnesia	22.0	30.80	30.20	28.67
Siliceous sand	0.5	0.10	....	0.13
	99.7	98.95	99.70	100.00

In my analysis a little peroxide of iron wrs present with the protocarbonate. On elevated ground east of the Folly River the ankerite is stated by Dr. Dawson to be decomposed to the depth of eight feet, and westward of the Acadia mine a shaft sunk on the course of the vein passed through more than forty feet of yellow ochre containing a few rounded masses and irregular layers of ankerite. The yellow ochre afforded Dr. Dawson

Peroxide of iron 74.52
Alumina 4.48
Carbonate of lime and magnesia . . .40
Silica and silicates 6.20
Water, mostly combined 14.40
                                                    ———
                                                    100.00

whence we see that the original earthy salts have been almost entirely removed ; a little clay has been added, the iron has become oxidised and in that state has combined with water. The ankerite is sometimes mixed with spathic iron (which in its pure state consists of protoxide of iron 62.07 and carbonic acid 37.93 parts in the hundred) and they are at the particular spot referred to decomposed to a much greater depth than usual. The oxidation of this would give rise to red ochre, consisting of peroxide of iron, and the absorption of water might give rise to another red ore called turgite, or to yellow and brown iron ores : when manganese ores are present various other tints will be produced which will be further modified by white or lighter coloured earthy minerals. An ore, probably ochrey red ore, gave Jackson :—

Peroxide of iron 70.20
Alumina 6.80
Carbonate of lime 5.60
Carbonate of magnesia 2.80
Silica 14.40
Oxide of manganese 0.40
                                                    ———
                                                    100.20

Mr. A. Ross, of Folly, has been engaged in raising these ores for pigments ; he informs me that the deposit of paint is situated on the Folly Mountain, eight miles from the Bay of Fundy, and about three-fourths of a mile east of the Folly River, and less than one-fourth of a mile north of the Intercolonial Railway. It consists of purple and yellow ochre, not in a regular bed but interspersed among rocks. A bright red is obtained by burning the yellow ochre. The most valuable is the purple which is worth about $5 a ton at the mine. The depth of deposit seems to range from five to twenty feet, but its extent is not known on account of its not having been sufficiently explored. The amount raised has been from five to forty tons a year. The purple ore has been made into a paint in Halifax, where I learned that by particular manipulation a very durable preservative for wood and iron is obtained: the paint stands well when exposed to friction, and adheres very firmly to dry, not green, wood. The red is only used to alter the tint of the purple.

Paints of Onslow, Colchester Co. The property of the Onslow East Mountain Manganese and Lime Company contains several materials suited for making paints and washes, some of them consisting of more or less ferruginous marls and other clayey mixtures One is soft, of a red colour, and when burned gives an agreeable pink wash; a second is yellow; a third cream coloured; one hill furnishes five distinct kinds of these substances. The most important of the minerals will probably be found to be an umber formed from the weathering of a ferruginous limestone; at about thirty feet from the base of one of the hills I saw a mass about two feet long and in some parts nine inches thick consisting of fine, soft, rich reddish brown umber upon a small nucleus of the rock; about twelve feet higher up the hill the same kind of umber was obtained on digging a little below the surface, hence there is probably a considerable deposit. A portion of the mineral was examined and found to be practically free from lime and to contain but a moderate amount of sand; from this the umber could readily be washed. The chief constituent is hydrated peroxide of iron, oxide of manganese is perhaps the next most abundant.

Paint of Chester Basin, Lunenburg Co. Rock found abundantly about 5 miles from Chester was at one time largely used in the manufacture of paint, by Mr. R. D. Clarke, now of Halifax, under the name of Petro-metallic Paint. I examined the locality in 1866 for W. Sutherland, Esq., and found the rock cropping out at several points over a considerable area of his property, and in its neighborhood; in some places its thickness was seen to be 4 or 5 feet, and old pits were pointed out from which it had formerly been taken at a depth of several feet. The exposed rock is of a rich brown colour and when fully decomposed furnishes a good and very soft umber which in some places is three or four feet thick. The original rock is very hard, of a deep blue colour somewhat lustrous, and smooth; it contains numerous very small specks of iron pyrites distributed pretty evenly throughout. It was found on analysis to consist of the carbonates of lime, protoxide of iron, oxide of manganese and magnesia, with bitumen or organic matter, pyrites, and sand. The umber consisted of hydrated peroxides of iron and manganese with *very small amounts* of lime and magnesia, sand was no doubt present to some extent. On a former occasion I had found about 20 per cent. of peroxide of manganese in a specimen of paint from the neighbourhood. The origin of the umber and of that of Onslow is similar to that of the ochres at Londonderry and a comparison of the quantitative analysis of these and of the rock from which they are derived detailed in describing them may be made in connection with the qualitative analysis just given of the umbers and the rock from which that of Chester comes. The changes consist, it will be observed, in oxidation of iron and manganese, absorption of water, and the removal of soluble salts.

The composition of the umber accounts for its value as a durable paint. I have tried it with oil and with turpentine and found it to work admirably. Mr. Clarke used to make paint also from the unweathered rock and considered it to be "stronger and better every way;" it was found to answer exceedingly well as ship-paint. While the prevailing colours are tints of red brown, the weathered rocks in the district afford also a yellow and a red ochre. At the local Exhibition of 1862 a shaving from a door coated with one of these paints 14 years before was shewn still remaining well covered, and also a piece of a cooper's adze from which the paint had not been worn by 10 years use, and it is well known that at Chester it has stood outside wooden buildings for forty years. It is esteemed to be fireproof and it no doubt affords effectual protection from the sparks and falling embers so dangerous to wooden structures during conflagrations. For cement made from the mother rock see *Limestones.*

Paints of Kentville, Kings County. Brown bog iron and manganese of different shades, and yellow ochre are found in abundance at Beech Hill and four hundred tons have been shipped to the U. S., within the last two or three years for the purpose of making paints and also for colouring glass.

Paints of other localities. Minerals suited for making pigments, as before stated, are found in many parts of the province: these are sometimes in considerable quantity at other places than those before mentioned. The following are the names of some localities furnishing ochres, umbers, or wad, but I am not in possession of details as to the amount found. Lochaber and Antigonish, Antigonish county; Chezzetcook and Jeddore, Halifax county; Petite Reviere and Bridgewater, Lunenburg county; Montegan, Digby county; and Louisburg and Sydney, Cape Breton. Washes are made from clayey minerals of Dartmouth.

Ochres sell here at about 7 or 8 dollars per ton when taken out; good umbers are worth $30 and even more in New York.

The beautiful blue phosphate of iron found near the surface at Antigonish, exists, unfortunately, only in small quantity.

CHAPTER VII.

ORES OF MANGANESE.

The existence of ores of manganese in the province has long been known, and several years ago a few barrels were sent to the United States but it was not till 1862 that mining operations on any scale were undertaken and since that time a considerable quantity of very fine ore has been raised and exported. The following account of the ores is taken chiefly from a paper of mine published in the 'Transactions of the N. S. Institute' for 1865, " On the ores of manganese and their uses," and from reports subsequently furnished by myself to the Teny Cape Co. and the Onslow East Mt. Co. after visiting their properties, and from information obtained directly from those engaged in operations at localities which I have not examined :—

The manganese ores found in quantity are pyrolusite, manganite, and wad, a notable amount of psilomelane is also met with, often mixed with pyrolusite ; all of these pass in commerce under the name of manganese.

Pyrolusite. " *Manganese.*" Pyrolusite, the mineralogical name of the most valuable ore known in the market as manganese, consists of the binoxide or peroxide of the metal and contains when pure 63.3 per cent. of the metal manganese and 36.7 per cent. of oxygen ; it is this oxide which gives value to all ores of manganese used in making bleaching powder (being called on this account available oxide,) and the amount of which is sought in the assay of manganese ores. Pyrolusite has metallic lustre, its colour is iron black, or dark steel grey, sometimes bluish, its powder is black, it is opaque and rather brittle, and is easily scratched with a knife. It occurs in several parts of the province, most abundantly, so far as is yet known, in the counties of Colchester and Hants which afford it in mining quantity.

Manganese of Hants Co. The limestones of Walton and Cheverie have long been known to contain the ore in irregular veins and nodules and it is from these places that small quantities were formerly shipped: from his examination of those rocks it was the opinion of Dr. Dawson that it seemed doubtful whether mining operations for manganese alone could be carried on without loss. In 1861, Mr. Nicholas Mosher, Junr., of Avondale, brought me samples from Teny Cape, on the Basin of Minas, about 5 miles from Walton, which I told him were good manganese ore. On diligent and continued search he found the ore to occur in nodules of all sizes, from that of a bean up to that of the lump of 24 lb. weight which was sent to the Exhibition of London, 1862, in a bed of earth about a foot thick, and a foot below the surface. In this mode of occurrence it was traced some 50 rods; then it was found in thin "veins" in the rock beneath which was a reddish limestone easily separable with a pick so as to expose sheets of the ore. In one place four veins were found in 10 feet, the largest being about 1½ inch in thickness. On digging 4 or 5 feet the ore was found to increase in quantity and then it became apparent that it was not in regular veins but in separate, often lenticular, masses in pockets, and it was so variable in amount that while on one occasion two and a half barrels were got by one man in a day, the average quantity obtained was about half a barrel per man per day. The first considerable collection of ore was landed at Windsor, in June 1863, for transmission to England. It consisted of 33 barrels, equal to about 7½ tons English; it was dressed ore and looked very uniform and rich in quality. A pupil of mine, Mr. D. Brown, had found in my laboratory 95 per cent. of peroxide in a sample from Teny Cape, and when this lot of ore was analysed in England it gave on the average 91 per cent. oxide and less than half of one per cent. iron; it sold there, half for £8.10, half for £9, sterling, per ton, to different purchasers. Messrs. Tennant, of Glasgow, are reported to have said (their consumption of manganese is vast) "they had never seen ore so fine." Results so encouraging could not fail to give an impulse to the Mosher Company's operations which were carried on with vigour, while others prospected in the neighbourhood and also began to work.

In April 1864, the ore shewed a thickness of 5 ft. 2 in., and in June I saw an enormous mass in place estimated to weigh some three tons. Mr. John Browne, the manager of the works, obligingly furnished me with a report dated Feb. 16, 1865, from which I give the following interesting extracts shewing the mode of occurrence of the ore and the amount then obtained. "On the south side of the ridge a large open cutting was brought in running nearly N. and S., in which was discovered the first large deposit at a depth of only 15 feet from the surface. It extended some 12 fathoms in length, varying in thickness from 14 feet to as little as 6 inches. From this pocket we took from 120 to 130 tons leaving nothing in the bottom but a few small veins. On these we sank a shaft, and at a depth of 15 feet, making in all 30 feet from the surface, we intersected pocket No. 2. immediately under the first and making in the same direction. The ore in the second pocket is of far superior quality to that found nearer the surface, and we have returned from it some 180 tons. Up to the present time we have been opening ground and prospecting. In conclusion I beg to state that our prospects are daily improving and I firmly believe that at no distant date the manganese mines of Teny Cape will hold a distinguished place in the list of *bona fide* and profitable mines of Nova Scotia."

In August, 1865, I visited the mine for Messrs. Mosher, Nash and Co., and reported on the improvements and the extension of the operations since my previous visit. The shaft had been sunk to a depth of 60 feet, and an adit had been driven, of about 290 feet, from the bottom of the hill to meet the shaft, affording an effectual drain. Good discoveries of ore had been made at nearly the lowest depth. At about 30 feet from the surface cuttings had been made east and west from the shaft, that on the east being some 100 feet; at 10 feet east I saw good ore in a soft rock, and it extended, Mr. Browne said, some 100 feet, imbedded in the same way, shewing a thickness of rock and ore of about 6 feet average; some 300 tons of ore had at that time been taken from the cutting. On the west side the cutting was about 36 feet, at about 5 feet from the shaft I saw a good exposure of ore from which a piece gave, when dried at 212°, 91.21 per cent. peroxide and a very small amount of iron. Near the mouth of the adit was a heap of some 2 or 3 tons of ore taken from a deposit two or three feet wide in a cross cut 12 feet north of a level running east and west of adit and at about 80 feet distance east of the adit, the rock carrying the ore was said to be 16 feet thick here and the ore is on the north wall but the level was driven on the south from the following of some small pockets of ore. As this deposit may be presumed to extend east and west like those at higher levels there was said to be probably a large amount of ore.

I had several pieces of the ore from this depth, 50 feet, brought me and selected some of them as representing the average of the whole; a good deal of barytes was present, especially on one of them, but when dried at 212°, an average selection gave 75.43 per cent. peroxide, and a piece weighing one-fourth of a pound, after dressing, gave, dried at 212°, 93.83 per cent. peroxide of manganese. The iron in the whole undressed sample was so small that it might be safely said to be in traces.

From the foregoing details it appears that the limestone contains the ore diffused through it in varying abundance in depth so far as the Mosher claim has been tried, and from what has been done to the west by Messrs. Weeks, and to the east by Messrs. Duvar and Co. in the same rock it is obvious that the ore is not limited to the breadth of that claim. I prospected the ground to the west adjoining the Mosher property for Messrs. Weeks and Ouseley, in June 1864, after examining some samples of ore which gave from 88 to 92½ per cent. peroxide, and found favourable indications in a hill some 60 feet in height. This hill ran east and west, and from 7 or 8 separate points on the north and south sides and at different elevations ores were take from these "veins" which had quite the appearance of those I had examined. Operations were afterwards commenced and during the rest of the year about 8 tons were raised which sold in Liverpool for £8 5s. per ton. Mr. O. Weeks informed me that 110 tons were taken out in all some of which sold for £8 15s. Immediately on the east of the Mosher claim Messrs. Duvar and Co. opened on a fine deposit of ore about 1865 and have exported a considerable quantity. The thickness of the limestone carrying the manganese is estimated by Mr. J. Browne to be 300 feet. The mines are about 1½ miles from a place of shipment.

It may be recorded as an interesting fact that one cargo of 120 tons of Teny Cape ore after being twice sunk in its passage from

H

the mines to England, via Windsor, finally sold in Liverpool, where analysis gave 89 per cent. oxide, at £8 5s. per ton. In his report on the Dublin International Exhibition of 1865, Dr Honeyman, Sec. to the Provincial Committee, said: "Mr. J. D Nash contributed an interesting specimen illustrative of our mineral wealth for which he has received an 'Honorable Mention.' This large mass of manganese" (it weighed about 3 cwt. and came from Teny Cape) "was unique on account of its size. The Zollverein exhibited specimens of manganese of considerable beauty and apparently equal to ours in quality; but they had nothing to compare with it in size. Its size and position in front of our court secured for it a sufficient degree of attention."

Explorations have proved the existence of rich ores of manganese in many parts of Hants County, some of the localities being near Teny Cape. At Walton, about five miles distant, in a westerly direction, seven barrels of ore were on one occasion got out in digging up a garden, and in the woods about two miles from Walton, a party of us saw fine ore in one place of which about 5 tons were afterwards taken out. Within the last year or so, Dr. L. Feuchtwanger, of New York, has undertaken the working of ores at Rainy Cove, or Pembroke, a few miles from Walton, and Mr. Ackerly, his manager, informed me in August that he had shipped about 7 tons of good ore. Specimens from Pembroke shewn me have quite the appearance of Teny Cape ore. Operations have not been continued through the winter, but will be resumed I believe in a short time: in all, probably, some 15 tons of good pyrolusite and a few tons of manganite have been raised. The ores have been found in detached pieces in the red clay and loam of the surface: this is quite analagous with the mode of occurrence at Teny Cape and Onslow, and it is thought that extensive deposits exist here. The ore is also found in conglomerate, in limestone, and in thin veins in sandstone as appears by recent explorations.

A certain quantity of manganite is generally found associated with these ores of Hants, but the amount is not large compared with the pyrolusite. Brown hematite is occasionally seen but it can generally be so easily detached by dressing that Mr. Browne thought he had not sent away ten pounds of iron ore in all the 500 tons he had at the time shipped from Teny Cape. At Cheverie, about 7 miles west of Walton, abundance of manganese is found of which the greater portion appears to be manganite, some of it however is pyrolusite. Hence we see these ores distributed east and west for about 15 miles, they chiefly are found in an impure limestone of lower carboniferous age, varying in composition, which is generally reddish, sometimes grey, and in parts contains a notable amount of magnesia; they are frequently associated with barytes. Some 10 or 12 miles to the south of this range many boulders of ore have been found at Douglas by Mr. Mosher, two of these weighed respectively 35 and 184 pounds. The lesser one of these was analyzed in my laboratory by Mr. H. S. Poole, who found it to contain:—

```
        Hygrometric water.....................1.660
        Water of composition..................3.630
        Peroxide of iron......................  .603
        Soluble baryta........................  .724
        Gangue; (barytes?) ...................1.728
        Oxygen (by loss)......................7.035
        Peroxide of manganese................84.620
                                            ─────────
                                             100.000
```

It is obviously a valuable ore consisting to a great extent of pyrolusite, the rest of the manganese ore is psilomelane. The carboniferous district here lying between the basin of Minas and the metamorphic rocks of Douglas and Rawdon is from 15 to 20 miles broad and at its northern edge, as seen above, manganese is found for 15 miles from east to west; so far as I know it has only been met with in place at one locality on the southern edge, but it will probably be found in all limestones underlying the plaster of Maitland, Newport, and Windsor. The locality I refer to is W. Lyons's, on the west arm of the Avon River, in Falmouth, a few miles from Windsor, where I found it in the side of a bank. This locality is about 30 miles south west of the Gore, Douglas, where Mr. Mosher found the boulders of which the analysis is given above.

The amount of ore shipped from Hants county, chiefly from the Teny Cape mines, is probably about 1200 tons English. Of that sent by Nash, Mosher and Co., the per centage of peroxide varied from 73 to 96, the average being about 80 per cent. by analysis made in England: it is worth observation that, as Mr. Nash, to whom I am indebted for much information as to quality and price of ores, tells me, a great difference was found when the assay was made on unground and on ground ore. Thus assay made on unground ore in Liverpool gave a result some 15 per cent. below that got from the same ore when ground altogether in London. Where barytes is present an unfortunate selection of a small specimen for analysis may give a very unfavorable opinion of a cargo of ore. As regards price it varied within a few pounds per ton, the highest price obtained being £9 10s. sterling per English ton. Mr. Nash has a standing offer of £6 12s. 6d. sterling per ton for ore of 70 per cent. and 2s. 6d. more for every additional unit per cent., the firm making the offer being willing to take 400 tons a month. By Boston dealers $30 a ton are given for the ore on the wharf, either at Windsor, or Teny Cape, it being valued at double the rate of New Brunswick ore of the same per centage. Dr. Feuchtwanger tells me that Nova Scotian (no doubt Teny Cape) ore sells at $30 a ton wholesale, and two cents a pound retail.

Manganese of Onslow, Colchester Co. Pyrolusite has been found in quantity sufficient to encourage a mining company whose object is to work it in connection with other minerals found on their property. The locality of the Onslow East Mountain Manganese and Lime Co.'s land is at Little's and Blair's farms about six miles from Truro. I examined the property in 1866 and give the following details from a report subsequently furnished by myself, and from reports by Mr. Barnes and Dr. Honeyman. The prevailing rocks are limestones which from the character of the fossils observed some short distance from Little's house I pronounced to be lower carboniferous. The manganese occurs in nodules in an impure red limestone forming a hill some 70 feet in height at the highest point and gradually sloping for about 600 yards north and south to the level of the brook at its base: from east to west the hill is probably 150 to 200 yards; the ore had been found by tunnelling 55 feet in from the east face of the hill and 65 feet north and south near the tunnel, this being about 30 rods from the northern end of the hill, also in the brook about 160 feet south of the tunnel and in a limestone on the opposite side of the brook, which runs at about 40 feet mean distance from the base of the hill, at about 500 feet south of the tunnel. On the occasion of my visit the ore was being taken out of the east face of the hill by two men; it was in the form of nodules, varying in size from that of an egg to pieces weighing between 2 and 3 cwt., embedded in the limestone. At that time there was got during the first day about 3 cwt. by picking and a single shot. There had previously been obtained in this way about 4 tons. On the second day the soil close to the base of the rock was tried and about 2 cwt. got, on the third day about the same amount. On the last day I observed what may be important; the general form of the nodules is distinctly rounded; but I noticed two perfectly flat pieces of ore, of perhaps 5 or 6 lb. weight; the form of these may indicate veins or beds, which are more continuous deposits than the pockets usually found. From what I saw I had no hesitation in saying that the promise as to quantity was good: in all, with the amount of about 7 cwt. got in the three days I was there by two men, some $4\frac{1}{2}$ tons had then been extracted. Since my visit, Mr. Flemming, who had charge of the operations, has prospected the hill in depth; within 10 or 12 feet of the surface he found small pockets of nodules weighing from $\frac{1}{2}$ a pound to 3 or 4 lb.; about 30 feet down the pockets contained nodules of from 5 to 25 lb. while at the level of the brook he found as many hundredweight in a single pocket as pounds near the surface.

As regards quality the ore is very good; some of it excellent: no dressing had been attempted when I saw it, the ore had simply been washed in the brook but even the large nodules looked of the finest quality, examination of samples from two different nodules on my return home showed that very little mineral but pure peroxide of manganese was present, the amount of iron was very small. Iron ore was present in some samples but could no doubt be kept very low by good dressing. The whole amount of barytes seen was remarkably small. Mr. Barnes reported that the appearance of the ore and its mode of occurrence are precisely similar to the same deposits at Teny Cape and believed that large and valuable deposits will be found. Dr. Honeyman reported that from the ore having been found at Onslow and at Salmon river about two miles distant it would probably be found throughout the entire length of the intervening limestones, and that, at the time of his visit, some 3 months subsequent to mine, some 5 or 6 tons of manganese were ready for exportation. The prospectus of the company states that only one barrel of second class ore could be got out of this quantity and I understood that £6 currency had been offered for the ore in the province.

Manganese of Amherst, Cumberland Co. During a tour made in 1861, in company with R. G. Haliburton, Esq., at that time Secretary to the Commissioners for the London Exhibition, for the purpose of explaining the objects of the Exhibition and urging the sending of minerals and other articles, a very fine specimen of manganese was put into my hands, as found near Amherst, by Mr. Boggs, the Agent of the General Mining Association, at the Joggins Coal Mine. It was very soft and fine grained and gave on analysis :—

> Water 0.61
> Peroxide of manganese 97.04
> Gangue and loss 2.35
>
> 100.00

these results shew that it consisted of nearly pure pyrolusite, its characters were those of the ore from Saxony which is very much esteemed from the ease with which it is ground. (On one occasion a somewhat similar variety of ore was found at Walton. I was informed that a dealer in the United States once sent for Saxony ore, to England I believe, and he was delighted with two barrels of Teny Cape ore which were sent him as from Saxony.)

Manganese at Fennerty's, Halifax Co. Pyrolusite occurs in pockets containing about a hundredweight of ore, giving some 70 per cent. peroxide, either in whin or granite, at Fennerty's, near Halifax, on the Windsor road, as I am informed by Mr. Mitchell, who says that some attention was given to it about 10 years ago.

Manganese at Greenwich, Kings Co. In apparently similar mode of occurrence an ore composed of pyrolusite and psilomelane is found at Greenwich, near Wolfville, where about a ton of ore was said to have been taken out about 1864. I visited the place in 1865 and saw evidence of mining operations having been prosecuted, but the place of extraction was covered up. The ore occurred in a hard siliceous rock probably quartzite, in slate.

Manganese, pyrolusite, is also found at Springville, near Pictou, in Pictou Co.; in Musquodoboit, Halifax Co.; and, mixed with siliceous pebbles, near the Cornwallis bridge, King's Co.

Manganite.—This ore of manganese differs from pyrolusite in many respects. Chemically it is the hydrated sesquioxide of the metal, containing:—

Water.................................10.21
Sesquioxide of manganese...............89.79

                                      100.00

This amount of sesquioxide is equal to in round numbers 49 per cent. of binoxide, peroxide, or available **oxide**; hence for such **purposes** as the making of bleaching powder this ore in its **purest** form can never be equal to even the inferior qualities of pyrolusite which often contain 50 and 60 per cent. of available oxide, but for certain purposes, as will be mentioned, it is valuable. Manganite is also very much harder than pyrolusite, its colour is the same, steel-grey and iron-black, its powder is reddish brown. As before mentioned it is frequently found associated with pyrolusite at Teny Cape and elsewhere, it is often crystallized on the other ore. It is abundant at Walton and Cheverie. At the former place I have picked it out of the heaps of stones in fields near the river, and a bed of it is said to crop out in the river bank near the bridge, this I did not see, no doubt from its being covered with dirt. It is found at Cheverie in nodules on the beach above highwater mark and has been dug on the upland less than two miles from the beach; it was formerly shipped, but to what extent I have not been able to ascertain. In 1863 about 50 tons are reported to have been sold in the United States. A specimen of the Cheverie ore gave me on analysis:—

Water..................................10.00
Sesquioxide of manganese...............86.81
Gangue, (silica and barytes)............1.14
Oxide of Iron, soluble baryta and loss.. 2.05

                                       100.00

Available oxide of manganese...........47.73

Uses of the foregoing Ores of Manganese.—These ores are employed for a great variety of purposes in certain arts and manufactures of a purely chemical character, or in which the aid of chemistry is necessary, and according to the application to be made of them they are required of different qualities. In most cases a rather high percentage of available oxide is necessary, and for certain uses there must be little else in the ore and especially iron must be either absent or present in very small amount. The ores are used chiefly in making bleaching powder, glass, pottery, iron and steel, and in dyeing and calico printing, in the preparation of manganates, and permanganates, and boiled oil; they have been recommended also as deodorizers and purifiers of water, and as cheap agents in the extraction of gold from quartz by a process which is carried on at several establishments on a small scale near Grass Valley, in California, with satisfactory results. The following is abridged from Dr. Calvert's account of his process:—

"It may be advantageous to persons interested in gold mining to be made acquainted with a new and simple method of extracting gold from its ores which presents the advantages of not only dispensing with the costly use of mercury, but also of extracting the silver and copper, as well as the gold, which the ore may contain. Further, it may be stated that the process can be profitably adopted in cases where the amount of gold is small, and the expense of mercury consequently too great. I propose the following plan for extracting the gold on a commercial scale:—The finely reduced quartz should be intimately mixed with about one per cent. of peroxide of manganese; and if salt be used it should be added at the same time as the manganese, in the proportion of three parts of salt to two of manganese. The whole should then be introduced into closed vats, having false bottoms upon which is laid a quantity of small branches covered with staw so as to prevent the reduced quartz from filling the holes in the false bottom. Muriatic acid should then be added if manganese alone is used, and diluted sulphuric acid if manganese and salt have been employed; and, after having left the whole in contact for twelve hours, water should be added so as to fill up the whole space between the false and true bottoms with fluid. This fluid should then be pumped up and allowed to percolate through the mass; and after this has been done several times, the fluid should be run off into separate vats for extracting the gold and copper it may contain. To effect this, old iron is placed in it to throw down the copper; and after this has been removed, the liquor is heated to drive away the excess of free chlorine, and a strong solution of green vitriol or copperas is added, which throws down the gold as metal. If silver is present in the ore, it is necessary to use the sulphuric acid and salt with the manganese, taking care to employ six parts of salt instead of three as above directed. The use of this salt is to take up the chloride of silver that may be formed, and blades of copper must be placed in the solutions to throw down the silver, then blades of iron to throw down the copper: the gold being extracted as before."

It is perhaps impossible to learn the total consumption of the ores for the various purposes to which they are applied; we know, however that Great Britain is the great seat of the chemical manufactures and that manganese ores not being raised there in sufficient quantity or of the requisite purity for all uses they are largely imported. The quantity produced in the kingdom is not more than perhaps three or four thousand tons a year at the most, and the ores are of an inferior quality. What facts I have been able to get as guides to the amount consumed in Britain and the United States are of great interest. The most extensive use of manganese is in the making of bleaching powders, chiefly chloride of lime. According to a report of Mr. Gething to the British Association, in 1863, the amount imported into the Tyne district alone for this purpose was given as 11,400 tons per annum. Although this district is a very considerable seat of chemical manufactures, there are other parts of the kingdom where very large quantities of manganese are required; the most important are Liverpool, the locality of Messrs. Mushpratt's, and Glasgow, of Messrs. Tennant's, gigantic chemical manufactories. Accordingly we find in the Statistics of the Alkali Trade of the United Kingdom for 1862, "that the annual consumption of manganese was then 33,000 tons for the manufactures depending on the products of the alkali trade, viz.: soap, glass, paper, cotton, linen, woollens, colours, and all chemical manufactures of any magnitude." This estimate, however, takes no account of the ore used in making iron and steel, and the quantity used for the making of bleaching powder has been increasing of late years partly owing to the use of grass and other materials in making paper. The London *Mechanics' Magazine* in drawing attention to the ores of manganese in this province as described in my paper in the Transactions of the Institute from which I have been extracting, said, in 1866,—"Apart from what is used in steel making and other branches of the iron manufacture, there are annually consumed in Great Britain somewhere between 40,000 and 50,000 tons of manganese ores." The amount used in the United States was given to me two or three years ago as from 500 to 1000 tons a year, and was considered to be probably on the increase. With regard to the quality of ores used for particular purposes and the prices which may be expected according to their purity and richness, some information has been already given and a few further remarks may be added. It is found that in making bleaching powder the ordinary ores, containing from 65 to 75 per cent. available oxide along with water, oxide of iron, carbonate of lime, barytes, etc., answer so well that the rich pure ores like those of Teny Cape are not bought for this use unless at a price far below that given by those who, like the flint-glass makers, require only such ores. One of the Messrs. Tennant, for example, said he could not afford to use Teny Cape ore; Spanish ore of some 70 per cent. could be bought two or three years ago for about £3 sterling a ton, and the 11,400 tons mentioned as used in the Tyne district were valued at £4 sterling; the considerable admixture of oxide of iron which these ores contain is of no consequence in this manufacture. It results from the existence of ores such as answer the purposes of the bleaching powder makers in Britain, Spain and Germany, that they will not give more than about £5 sterling for the rich pure ores of Teny Cape. It is from the glass makers especially who require the manganese for removing the colour imparted to glass by iron that the high prices before mentioned, (sometimes amounting to £9 10s. sterling) have been obtained. To them, as the London *Mechanics' Magazine* in the notice before referred to is careful to point out, "these Nova Scotian ores, so much freer from iron than any yet found in Europe, will be a great boon." These ores are also used in the making of fine pottery. The demand for these ores would not of course be equal to that for those less pure, but probably some hundreds of tons a year might be called for. If in the glass works to be established in the province the manufacture of flint-glass should form a part of the operations, the existence of those ores close at hand would be a great advantage. As an illustration of the way in which the ores are used in this manufacture and also in that of inferior glass I may state that Mr. Hobbs, of Boston, who has had a good deal of experience in the use of these ores from New Brunswick and Nova Scotia, has the ore washed clean at the mine, and sent to Boston. It is there sorted into three good qualities and refuse; the former are ground in separate mills till fine as flour, put in barrels papered inside, and the contents of each are assayed and sold according to assay. The first quality, free? of iron and containing about 98 per cent. oxide, is used for making the finest flint-glass. The second quality, also no doubt pretty free from iron, contains about 75 or 80 per cent. oxide ; it is used for making white phials. The third, of about 70 per cent. oxide, is employed for common glass bottles ; while the refuse, containing perhaps 25 or 30 per cent. iron, is used in making clear amber-coloured bottles for brandy, etc., and carboys. That there is always a demand for the ores fit for making bleaching powder is not only evident from the amount consumed but from the patents taken out for the purpose of restoring to its original state the oxide chemically altered in the manufacture ; one is recommended by its owner as restoring 52 per cent. oxide and as being capable of bringing the amount up to 70 per cent. which is known to be a very moderate per centage in these ores in this province.

With regard to the other applications of manganese the making of iron and steel is the most important. Even in small quantity the presence of manganese renders iron tough and steel better and more durable by removing silicon. It appears also to be very valuable in counteracting the bad influence of phosphorus in making iron brittle. As this is important in view of some of the iron ores in the western part of the province, at Nictau namely, containing phosphorus, it will be useful to give an abridgement of some statements of Dr. Crace Calvert on this subject. "Eight or nine years ago I observed that if manganese had not the property of removing phosphorus from iron it had that of hiding that element : in fact, I found that cast iron, containing as much as one or two per cent. of phosphorus would yield good mercantile iron if the pig iron contained at the same time 5 or 6 per cent. of manganese ; and I have lately heard that manganiferous ores have been used with great advantage by the Cleveland iron smelters to overcome the "cold shortness" of their cast iron, due to the presence of phosphorus." (Cantor Lectures, Chemical News, XIII. 94.)

Considerable quantities of manganese may be required if the alloys recently announced as made in Germany, realize the expectations naturally formed in the following description given of their properties. M. E. Prieger has commercially prepared alloys of manganese with iron, copper and tin, possessing valuable properties and the applications of which are constantly improving in number and utility. Of these alloys the most important are those containing 66.3 per cent. and 79.7 per cent. of manganese. Both are harder than tempered steel ; they are capable of receiving a very high polish, they melt at a red heat and can be easily poured; they do not oxidise in the air and even in water only superficially: their white colour is of a shade between steel and silver. Alloys of copper and manganese resemble bronze but are much harder and more durable. Alloys of tin and manganese are very fusible, durable, and easy to work; in colour and brilliancy they may be compared to silver. The iron and manganese alloy furnishes a very simple means of adding to iron or steel a given amount of manganese; by the addition of from 1.10 to 5 per cent. very satisfactory results are obtained.

For such purposes as the making of iron, steel, and alloys, the manganite of Cheverie would be found quite as valuable as the pyrolusite of Teny Cape and other places. I was informed by a gentleman in the United States that for a certain secret process the hard ore, manganite, was preferred, and understood that the 50 tons before mentioned as exported from Cheverie were used in the application; it was hoped that the demand would increase. The process, I am of opinion, had reference to iron or steel manufacture. In this province where the iron and manganese ores both abound it is obvious than an excellent opportunity exists for such applications as the preceding. Now that the Windsor and Annapolis railway is rapidly advancing to completion the manganese ores of Hants county will be brought within practical reach of the iron ores of Annapolis county and the smelting houses of Clementsport and Nictau now lying idle may derive the full benefit of the advantages resulting from the use of manganese in making iron, steel, and Prieger's alloys. Since ores of manganese are not found to any useful extent in Ontario or Quebec (Geology of Canada, p. 751) those of this province may be called for there in any such metallic manufacture as those in question. The extensive works just commencing at Montreal primarily for making iron and steel from titaniferous iron sand under a patent taken out by the director, Mr. F. Ellershausen, well known in this province for his enterprise and energy, every effort is likely to be made to produce the best results possible from the application of the latest fruits of scientific research

A few words on the manganese of other countries may be found interesting. The ores of New Brunskwick, according to Professor Bailey, (Observations on Geology of N. B., 1865.) occur at the base of the lower carboniferous veins and in the overlying new red sandstone. In 1864, the known amount of ore raised and mostly sold was given as 1250 tons, and a large quantity must have been formerly used in chemical works at Shepody. In a report to the British Association, in 1863, it was stated that the richest ores were at that time mostly imported to Britain from Spain, where they are found in schistose rocks, sometimes 800 feet in height: they are sometimes found in pockets when they are quarried by picks, and occasionally by blasting. The quality of the ore varies from 50 to 90 per cent. peroxide, and to obtain the richer ore men and boys are employed to break and sort it; when ready it is put into sacks and carried a distance of 20 to 35 miles on mules' backs to ports of shipment on the Mediterranean. The first cargo was imported to the Tyne, in 1857, by Mr. Gething, the author of the report. Mr. Outram, Junr., informed me that the Spanish ores reached England as ballast. In Vermont, Dr. W. H. Weeks saw the manganese being taken out of a gravel bank in very small pieces, from the size of a pea to that of an onion, compact, very black and not shewing crystals. Iron ore, said to be very pure, was taken out of the deposit, and also ferruginous ochre. The process of getting and cleaning the manganese is slow and must be expensive, the ore is washed in pans very much as gold is washed here. The quantity is small in proportion to the amount of material operated on, compared with that of Teny Cape. The ore is pure when thoroughly cleansed, but is not easily got so as the iron adheres very tenaciously.

The advantages belonging to the ores of Nova Scotia as regards quality, facility of extraction, dressing, transport, and shipment, will be remembered.

Wad.—This is the name sometimes given to earthy and bog manganese ores, which are of various colours, as brown, chocolate and black, and exist as powders, and in lumps or nodules which are either very soft and easily crushed in the hand, or of a moderate degree of hardness. They consist of the hydrated oxides of manganese and iron with other substances, such as clay, limestone, and often a little oxide of cobalt. Some of these earthy ores are described under the head of paints, for which purpose they are somewhat extensively employed. In the brown paint of Bridgewater, Lunenburg county, I found 11 per cent. peroxide of manganese; in that of Chester Basin, Lunenburg county, about 20 per cent.; in a black wad from another locality, unknown to me, I have found 56 per cent., and a little cobalt: black wad, was brought from near Parrsboro, Cumberland county, by the Marquis of Normanby, when here as Lieutenant Governor, and sent to me for examination through Colonel Nelson, who, however, so far as I recollect, did not state in what quantity it was found. With regard to the value of these it is said by Mr. Outram, Junr., to be useless to send them to England unless they contain upwards of 65 per cent available oxide. Mr. Outram, Senr., formerly shipped brown wad from near Jeddore to England, but found it not remunerative. Such ores, however, appear not only to be valued as paints in the United States, some of them selling here, as mentioned under that head, for exportation to Boston, at about 7 dollars a ton, but Dr. Feuchtwanger, of New York, informed me that wad of about 50 per cent. available oxide would probably fetch some $15 a ton in that city, even though containing 12 per cent. or so of peroxide of iron. I think it quite probable this would be found useful in making brown glass, such as mentioned in speaking of the application of pyrolusite.

CHAPTER VIII.

GYPSUM—ANHYDRITE—BORATES—BRINE SPRINGS—SALT—MAGNESIA ALUM.

Gypsum and Anhydrite. These "minerals" are so closely associated that it will be convenient to consider them together. They are both known locally as plaster, the former being called soft, the latter hard; in the official returns they are both classed under the name of gypsum and, when ground, as plaster. Gypsum is composed of sulphate of lime and water; the pure mineral contains:—

$$\begin{aligned}
&\text{Lime} \ldots \ldots \ldots \ldots \ldots \ldots \ldots 32.55 \\
&\text{Sulphuric acid} \ldots \ldots \ldots \ldots 46.51 \\
&\text{Water} \ldots \ldots \ldots \ldots \ldots \ldots \ldots 20.94 \\
&\phantom{\text{Water} \ldots \ldots \ldots \ldots \ldots \ldots} \overline{100.00}
\end{aligned}$$

Anhydrite contains simply sulphate of lime, the absence of water being the circumstance indicated in the name; the pure mineral is composed of:—

$$\begin{aligned}
&\text{Lime} \ldots \ldots \ldots \ldots \ldots \ldots \ldots 41.18 \\
&\text{Sulphuric acid} \ldots \ldots \ldots \ldots 58.82 \\
&\phantom{\text{Sulphuric acid} \ldots \ldots \ldots} \overline{100.00}
\end{aligned}$$

These minerals occur together in alternating beds or masses forming immense deposits which have long been proved to be of great economic importance; these occur exclusively in the lower carboniferous series in close association with the sedimentary limestones found in enormous quantities in the north eastern parts of the province including Cape Breton. Geological details of the most interesting nature respecting these rocks will be found in Acadian Geology. In comparatively small and often in absolutely small amount gypsum occurs in the fibrous state in the trap rocks especially about Blomidon, and I have seen beautiful specimens of selenite in the trap of Two Islands. The exposure of the gypsum beds is often on a grand scale; a few miles from Windsor on the Newport road lofty white cliffs, chiefly of anhydrite, are seen, and in the same county, Hants, on the Shubenacadie, the Big Rock at one time presented a snowy front of gypsum nearly 100 feet high, and again, close to Walton, are white plaster cliffs of considerable height. At Ogden's Lake, Antigonish Co., is a beautiful cliff of white crystalline gypsum, 200 ft. in height, fronting the sea at St. George's Bay. Dr. Dawson also describes the bed of gypsum at Plaster Cove, on the Strait of Canseau, Cape Breton, as 50 yards thick, and as well exposed at the head of the Cove in a cliff of 80 feet in height. Near the mouth of the Mabou river, Cape Breton, there is another enormous bed of gypsum, which was being quarried when Dr. Dawson visited it, for the purpose of making road embankments, no other rock being available at the spot: very large pits had been excavated in the outcrop of the mass; one of them formed a grassy amphitheatre capable of containing hundreds of persons. These examples will suffice to shew the very extensive development in which these beds are to be seen, while the following tables convey important information on the economic value of the minerals. From the census it appears that the amount of gypsum *quarried* was returned as follows:

GYPSUM QUARRIED IN NOVA SCOTIA.

1851	1861
79,795 Tons.	126,400 Tons.

The returns for the latter year shew that it was quarried in eleven out of the eighteen counties in the following quantities and also give its value:—

GYPSUM QUARRIED IN 1860.

Counties.	Tons.	Value.	Counties.	Tons.	Value.
Colchester.	6,026	$5,407	Lunenburg.	300	$120
Kings.....	0	0	Yarmouth.	0	0
Cumberland	259	206	Digby.....	0	0
Annapolis.	0	0	Guysboro..	250	190
Pictou.....	70	46	Victoria....	0	0
Hants....	118,215	77,883	Queens....	0	0
Sydney....	10	10	Shelburne..	0	0
Inverness..	12	21	Richmond.	1,470	1,227
Halifax....	58	53	Cape Breton	30	24
Total......................				126,400	85,076

No census having been made since 1861 we have official details only with regard to the exportation of the rock, these are given separate from those of other articles in the Trade Returns for the following years only. In the table below the values are all taken from these records; in four years, as indicated, the amount is estimated from the probable average value. In the first three cases the years end 31st Dec., subsequently the year ends on 30th September, the change being made in 1857 ; the returns for that year for the whole province were not printed, hence I give only the quantity exported from Windsor, and that for nine months only, and in 1867 the returns are for nine months for all the province except Hants county, for which they are nearly complete.

QUANTITY AND VALUE OF GYPSUM EXPORTED FROM NOVA SCOTIA.

Year,	Tons of 2240 lb.	Value in dollars.
1854	87,283	74,935
1855	95,301	80,875
1856	72,210	61,485
1857 (Windsor only, 9 mos.)	33,862	11,050
1858 (Estimated)	86,291	69,015
1859 (Estimated)	109,243	87,395
1860	105,431	85,936
1861 (Estimated)	51,013	40,811
1862 (Estimated)	38,031	30,425
1863	46,739	30,625
1864	58,601	43,167
1865	56,155	45,088
1866	77,091	63,611
1867 (Partial)	103,426	88,486
Totals	1,020,577	813,904

Considerable insight into the gypsum trade is given by the following table compiled from the vastly improved Trade Returns of late years from which we learn the amount of plaster exported to different countries from various ports of the province. For the year 1867 the returns are partial as stated with reference to the preceding table. The rock shipped from Halifax is quarried at Maitland, in Hants county, whence it is transported by rail 39 miles.

GYPSUM.

Countries to which exported and Ports whence shipped.	Amount of Gypsum exported from Nova Scotia in the years ending 30th of September.			
	1864.	1865.	1866.	1867.
	Tons.	Tons.	Tons.	Tons.
To CANADA—				
From Antigonish	283	486	300	260
,, Arichat	790	985	712	560
,, Halifax			5	
,, Port Hawkesbury				277
To NEWFOUNDLAND—				
From Halifax			7	
To NEW BRUNSWICK—				
From Arichat	140			
,, Antigonish	54		54	
,, Horton	80		223	
,, Little River	70			
,, Maitland (ground in bbl.)	(185 bbls.)	(35 bbls.)		
,, Pictou			45	
,, Walton		1,045		100
,, Yarmouth	30			
,, Pugwash		367		
To PRINCE EDWARD ISLAND—				
From Antigonish	505	716	388	85
,, Halifax				10
,, Pugwash	348	232	42	826
,, Pictou	196			
,, Port Hawkesbury			30	79
,, Port Mulgrave		30		36
,, Port Hood	158		115	30
,, Little River			50	
,, North Sydney			75	
,, Wallace			60	
,, St. Ann's				8
To BRITISH WEST INDIES—				
From Halifax			(65 p'kges)	
To ST. PIERRE—				
From Halifax				40
To UNITED STATES—				
From Arichat	440	685	410	636
,, Amherst				8
,, Five Islands		60	120	
,, Halifax			1,683	835
,, Horton	80	85	65	
,, Port Greville				100
,, Londonderry	550	390	1,775	545
,, Ratchford's River			100	
,, Parrsboro'	1,050	475	585	820
,, Truro		655	230	
,, Cornwallis			400	711
,, Cheverie, Hants Co	8,220	8,130	14,104	13,395
,, Hantsport, ,,	5,130	5,845	5,210	7,735
,, Maitland, ,,	4,697	3,130	3,855	1,555
,, Walton, ,,	1,450	1,375	3,475	7,400
,, Windsor, ,,	34,254	31,464	42,963	67,350
	58,625	56,155	77,091	103,426

GYPSUM. 131

From the foregoing table it appears that Hants is the chief gypsum raising county, and Windsor its principal port of shipment. In fact by far the largest quantity of the rock is quarried at Windsor or in its neighbourhood where operations have been carried on for some 80 or 90 years. The quantities exported in former years were very large as they are now, as seen in the following table shewing the

AMOUNT AND VALUE OF GYPSUM EXPORTED FROM WINDSOR.

Year.	Tons of 2240 lb.	Value in Dollars.	Price per ton in cents.
1833	52,460	32,785	62
1834	48,710	31,180	64
1835	36,680	22,925	62
1836	47,935	29,955	62
1837	34,649	21,655	62
1838	36,422	24,465	67
1839	39,491	24,805	62
1840	48,693	30,430	62
1841	35,878	22,940	63
1842	38,450	24,030	62
1843	28,310	17,070	60
1844	29,570	18,535	66
1845	35,764	20,910	58
1846	37,048	23,750	62
1847	20,472	11,960	58
1848	38,423	24,090	62
1849	32,512	20,315	62
1850	30,661	19,165	62
1851	32,283	20,175	62
1852	43,870	29,065	66
1853	55,838	45,745	82
1854	48,268	41,070	85
1855	50,015	46,110	92
1856	34,098	33,345	98
1857	40,637	36,965	91
1858	47,835	42,265	88
1859	60,434	59,955	98
1860	61,110	52,580	86
1861	22,162	13,754	62
1862	21,902	13,141	59
1863	34,204	23,912	69
1864	29,437	22,502	76
1865	30,749	26,136	85
1866	55,751	49,256	88
1867	63,655	54,213	85
Totals.	1,404,376	1,031,154	73 { Average of whole.

On the wharfs of this port are collected the products of some 7 or 8 quarries, the most remote being those of Newport distant some 6 miles along the line of railway (to Halifax) by which the plaster is brought in. The years of the late American war were a season of great depression, latterly however very great activity has prevailed; there were

CLEARED FROM THE COUNTY OF HANTS, N. S.
From January 1st, to Dec. 31st, 1867.

Ports.	Tons of 2240 lb.	Value in Dollars.
Hantsport	9420	9112
Maitland (9 months)	2440	1708
Walton	9845	7384
Cheverie (chiefly hard plaster)	14,799	8190
Windsor	63,655	54,106
Totals	100,159	80,500

The exportation would have been much larger but for the closing of navigation about a month earlier than usual. It is stated that the amount of plaster stone sent from France is "very large"; it was, in 1859, upwards of 6915 tons. This would not be thought even a large quantity from a port third in rank in Hants county. Hunt's Guide to the Museum of Practical Geology also gives the amount of gypsum imported into England as 1781 tons and of Plaster of Paris 5,155 tons, (probably ground,) the aggregate consumption being reckoned at 30,000 tons (valued at £10,000 sterling,) chiefly used in moulding at the potteries.

The price of gypsum at Windsor last year was from 90 cents to a dollar a ton shipped: at many other places different prices obtain, generally from local causes. The price at the Grand River, the only part of the old "Canada" where workable deposits occur is about two dollars a ton at the mine. The thickest deposit is about 7 feet only and the amount raised was in 1863 given as 14,000 tons. The deposits are quite insignificant compared with those of Nova Scotia. The low value here per ton, as remarked by Dr. Dawson, shews the facility with which gypsum can be raised in a country where the price of labour is by no means low. Ordinary quarrymen receive at Windsor a dollar a day, those who conduct the blasting $1.25 or $1.50. The average selling price in New York is $3.25, in gold, per ton of ordinary gypsum, exceptional qualities occasionally bring much higher prices; it is bought here by the British ton of 2240 lb. **and sold in the United States no doubt by** the American ton of 2000 pounds.

In the district including a few miles about Windsor the quarries are worked on parallel beds running E. and W., the most northerly extending from Windsor through Wentworth and Newport, and probably continuing in unbroken series in depth to Maitland, 80 miles to the east, where plaster is also worked, and even beyond. The distance across the strike from the north at Windsor to the most southerly quarries is about three miles, at Windsor the dip is gently to the south. The largest quantity of gypsum is now raised at the Clifton quarry, formerly the property of Judge Haliburton, now owned by Mr. Pellow, close to the town of Windsor, where operations have been carried on some 40 years. The principal rock is gypsum, the anhydrite or hard plaster being found in lenticular masses from 2 to 10 feet thick in the centre and sometimes fifty feet long imbedded as it were in the soft plaster. Mr. Pellow considers that the amount of plaster got here has varied for the last 30 years from 10,000 to 30,000 tons a year, and for the last 10 or 12 years from 20,000 to 30,000 tons. The quarry is now roughly estimated as being 800 feet long, 150 ft. broad, and on an average 40 ft. deep. To the north the rock cropped out near the surface and to the south is now a face of some 30 ft. of plaster with a little limestone here and there. At the east end of the quarry a face fully this height is now to be seen on the east and north sides also. Operations having now been carried as far in depth as allowed by the natural drainage (of chiefly surface water) into the neighbouring river, a steam pump is erected by which the workings into the unknown thickness of the beds will be carried on in the approaching summer. On another range of beds to the south are extensive quarries, owned respectively by Messrs. Wilkins, M'Latchey, and Pellow, situated about 1¼ mile from Windsor, Mr. Wilkins's quarry is close to the line of railway to the town. The rock here is the best soft blue, very free working, and a face can be got of from 15 to 40 feet. Mr. Pellow has 150 acres here nearly all plaster, he has traced the beds across the strike for 300 feet, he estimates that before the quarry came into his possession 100,000 tons of plaster had been removed. At some parts of this deposit a good deal of transparent selenite, or isinglass of the quarrymen, is found.

On the last range south are the quarries of Mr. Black where soft blue and white and also hard plaster are obtained; directly south of these are the metamorphic rocks of the Ardoise Hills.

At the Three Mile Plains, east of Windsor, is a very fine quality of soft white plaster, with some hard of a bluish colour. The rocks here do not seem to run in distinct beds but so far as exposed appear to be rather in separate masses extending over some 50 acres, Mr. Pellow further describes some of these as 50 feet in height with no soil on them ; a railway is now being made in from the station to a face of rock in Mr. Pellow's quarry about 30 feet in height from which about 1200 tons have been blown.

From the Wentworth quarries, about two miles from Windsor, about 40,000 tons have been raised the last two years. In February 1868 there were piled at the "Creek" 9,000 tons for exportation on the opening of navigation.

The great distinctions made in the export trade are between hard and soft plaster, and as regards the latter, between blue and white. The blue is furnished principally at Windsor and the immediate vicinity ; it is the kind chiefly used for agricultural purposes ; probably the greater part of the large amount returned as exported is so employed ; it goes to Boston, Portland, New York, Philadelphia, and Richmond. The greatest consumption appears to be in Virginia, Maryland, and Pennsylvania, where the ground plaster is applied as manure for tobacco and Indian corn ; before the late war it was beginning to come into favour with cotton growers and large orders were sent to the Northern States, but the war interfered with the attempts to use it and as yet I believe the trial of its merits has not been made to any extent. A considerable quantity is taken, Dr. Lawson tells me, from Portland to Upper Canada where it is ground and sold for top dressing at not less than a dollar a barrel. The white plaster, which is found chiefly perhaps, in this district, at Newport, Wentworth, and Falmouth where a very fine quality is met with, and now at Three Mile Plains, is sought for calcining. By this process, which consists of boiling or burning, the water belonging to the mineral is expelled and the powdered residue mixed with water becomes the plaster which is used in very large quantities here as well as abroad in finishing walls and ceilings. Other well known applications are the making of casts, models, copies of valuable statuary, and of large and expensive fossils, of which an interesting illustration is afforded in the restoration of the extinct animals in the gardens of Sydenham, and in some museums. For particular purposes, as the making of fine ceilings, and their centrepieces and cornices, a very white plaster is required. Of course the best qualities of gypsum for such purposes are the whitest, and of the varieties found the purest is the selenite which is tolerably abundant in some quarries and is often as transparent and colourless as the finest flint-glass. A few years ago a cargo of selenite was taken hence to the United States, doubtless for such a purpose, and five dollars a ton were paid for it here. It has recently been announced that gypsum can be used instead of china-clay in the making of paper. The whitest varieties would no doubt answer best.

It is obvious that the manufacture of all the cements of which gypsum is the basis could be carried on to the greatest advantage here so far as a perfectly inexhaustible supply of material of different qualities is concerned. These cements, such as Parian, Keene's, Martin's and Keating's, differ from common plaster of Paris inasmuch as they become much harder. It is well known that common plaster (of Paris so called) is got by calcining gypsum. The calcining is effected by boiling or burning, the former process being employed when the best plaster is required; it is called boiling from the circumstance of the gypsum undergoing a peculiar agitation (like that of water when boiling) when the water is escaping from the gypsum heated in a caldron. The burning, which is the process followed here, consists in building up lumps of gypsum into a heap with cord-wood distributed through it, and keeping up a very moderate fire: the burnt gypsum when cold is beaten to powder with a hammer, and, mixed with water, forms the plaster which is used for walls and ceilings. The proper temperature for making the plaster which most rapidly becomes hardest with the proper amount of water is 500° Fahrenheit, if the heat attains redness the gypsum becomes much more dense but does not set with water. Now if a rather small percentage of certain salts, such as sulphate of potass, alum, or borax, is added most important qualities are gained: the gypsum so treated will endure a red heat without losing its power of setting with water. It becomes much more dense than common plaster, and when mixed with water sets in a few hours and becomes so hard that it will take a fine polish. The cements called by the names before mentioned are prepared by these processes. Keene specifies his process to be as follows. One pound of alum is dissolved in a gallon of water, this solution is used for soaking 84 pounds of calcined gypsum in small lumps. These are exposed to the air for 8 days and afterwards calcined at a dull red heat. They are then ground and sifted. The fine powder thus produced is mixed with water into a paste which may be used as ordinary plaster; upon setting it forms a compact and durable body which can be polished or coloured without difficulty.— If half a pound of common copperas be added to the solution of alum the resulting paste has a fine cream colour and the hardened mass is said to resist the action of the air. Borax gives the Parian cement. Martin's cement is formed by combining pearl ashes and alum with the plaster, muriatic acid being sometimes added to prevent an alkaline reaction. Scagliola differs from these cements in being gypsum mixed with glue and then painted. Fictile ivory is made by dried plaster casts being allowed to absorb melted spermaceti wax and stearine. When the cast has drained, and before it has cooled, the superfluous wax is brushed off, and when quite cold the surface is polished by rubbing with a tuft of cotton wool. Stucco is coloured plaster mixed with size. Another valuable substance is obtained by mixing gypsum with a certain amount of water and then soaking it in hot pitch, it then parts with some water and takes up pitch and becomes so hard and susceptible of polish that it can be made into a variety of useful and ornamental articles. What success would attend attempts to make artificial stone fit for outdoor use from gypsum by action of soluble glass or silicate of soda remains to be seen. Some application of this kind (hardening gypsum) is understood to be in view in connection with orders that have been received here for large blocks of gypsum from the United States; the exact nature of the process or of the material used has not transpired. No objection was made to the high prices it would be necessary to charge for quarrying blocks of the required dimensions.

The compact white gypsum called alabaster is found abundantly in Antigonish county, and a very fine quality has lately been met with at Three Mile Plains, near Windsor. It is very suitable for indoor carved work, as was shewn by a small specimen from Antigonish county, carved by the late Mr. Harding, sent to the Dublin and Paris Exhibitions. It is very durable, as pointed out in another place. (Trans. N. S. Institute 1866, p. 65.)

The "isinglass" of the quarrymen, selenite of mineralogists, was mentioned as abundant in some quarries. It is so on the second range south of Windsor, and it was from the Retreat quarries very near Windsor that the cargo was taken at $5 a ton. It has been used as a material for filling fire-proof safes.

Between the several kinds of gypsum there is a difference of composition from the admixture of various amounts of carbonates of lime and magnesia, oxide of iron, clay, and silica. These ingredients are probably quite small in amount in most of the gypsums here, the quantity of some must be considerable, however, in a few of the deposits: these impurities do not interfere with the use of the rock as a manure, but some of them of course are detrimental in reference to making the whitest plaster. In certain applications, however, hardness rather than whiteness is the quality sought in plaster as shewn in the artificial stones and cements just described. It is said that the French plaster acquires a greater degree of solidity than any other known in Europe, and that this property is due to its containing about 12 per cent. of carbonate of lime. Gay Lussac says that the purest plasters are those which harden least; he does not, however, consider this to be dependent upon the presence of the carbonate of lime but upon the original hardness of the stone. I have examined several samples of gypsum from the neighboreood of Windsor and Newport, qualitatively, with the results named above, those of grey and blue colour contained a considerable amount of foreign ingredients, chiefly carbonates, and in very white specimens I found notable quantities of carbonate of lime and magnesia in some cases. The hardness is very variable. An opinion seems to prevail that "rotten plaster," or that which has been exposed to the weather and become a crumbling mass has lost its "strength." I analysed such a gypsum on the property of Mr. O. King, at Windsor, and found it to contain:—

Water and trace of carbonic acid.....21.16
Lime.............................33.02
Sulphuric acid....................45.99
                                 ——————
                                  100.17

or almost exactly the quantities of constituents proper to pure gypsum: hence the rock was entirely unchanged, chemically, by exposure, and was fit for all the purposes to which gypsum can be applied. Of course in the crumbling state it could not be shipped to advantage, but for local use as ground plaster, or for "boiling," the condition to which it has been brought, chiefly no doubt by the action of frost, would be advantageous.

Gypsum is sometimes nearly black; I analysed a specimen of this sort, from a deposit near Walton, and found it to contain:—

    Gypsum........................80.45
    Anhydrite...................... 2.84
    Bituminous matter............... 1.53
    Sand and clay................... 7.94
    Carbonates of lime and magnesia, with
        alumina and oxide of iron......} 7.23
                                     100.00

With regard to gypsum varying very much in hardness, I examined a specimen of white colour from Falmouth, which was much harder then some varieties; on analysis it gave:—

    Water 20.94
    Sulphate of lime, by loss............79.06
                                     100.00

hence it was really gypsum, it should according to Gay Lussac give after burning good hard plaster.

Anhydrite (the composition and mode of occurrence of which have before been given), or dry sulphate of lime, is of various colours, as dark blue, purple, and grey; on exposure to the weather it becomes white with a peculiar roughness of surface which has obtained for it the name of sharkstone. Known as hard plaster it is always much harder than gypsum but varies very much in its hardness as a rock. Some samples give a very clear sharp sound under the hammer, others sound dull. True anhydrite can give no water, but hard plaster is often a mixture of anhydrite and gypsum, and then of course will afford water. Hard plaster is used at Windsor as a building stone, viz: for the foundations of houses and walls to support wooden fences. It makes apparently a good substitute for marble in indoor work; at the Paris Exhibition were shewn a small table top and a pedestal made and polished in Windsor by Mr. Woods: the latter especially was much admired at the preliminary local Exhibition. How far the beauty of the surface will be retained on exposure, or by what means it could be secured if not permanent without some aid, remains to be proved. Blocks of almost any required dimensions could be obtained.

Anhydrite does not admit of being made into plaster by burning but it is equally good with gypsum for agricultural purposes, in fact so far as its chemical composition goes it is some 21 per cent. more valuable as it is free from this amount of water held by gypsum. On account of its hardness it cannot be ground in gypsum mills but is crushed by stamping. The plaster from Cheverie, Hants Co., is chiefly of this kind and is mostly shipped to Bridgeport, near New York, where it is almost the only sort employed. Its value at Cheverie is about 55 cents a ton.

It is worth notice that probably anhydrite would answer as well as gypsum, instead of porcelain clay in the manufacture of paper.

Borates and other Minerals found in Gypsum and Anhydrite. No considerable amount of foreign minerals has been found in the deposits just described, but small quantities of various kinds have been met with from time to time which are very interesting from a scientific point of view and some of which will prove very valuable if found in abundance. Details respecting some of these minerals will be found in papers of mine contributed to the Nova Scotian Institute of Natural Science, Nov. 4th, 1867, and the *Philosophical Magazine*, Jan. 1868, where references are made to former papers describing others of them, and from these I give an abridged account of such minerals as have so far been met with, here shewn in a tabular form:—

Hants Co., Nova Scotia, has deposits made up of	Gypsum containing	Natroborocalcite, cryptomorphite, silicoborocalcite, glauber salt, common salt, arragonite, calcite, and selenite, as distinct accessory minerals, and also, to be found on analysis, carbonates, partly of magnesia and protoxide of iron, peroxide of iron, bituminous and carbonaceous matters, clay, and a very small quantity of silica.
	and	
	Anhydrite containing	Silicoborocalcite, selenite, and arragonite or calcite, as accessory minerals, and also, to be found on analysis, carbonates, partly of magnesia, and a very little silica.

of the minerals here named the three first mentioned are those which will prove valuable if found in quantity. They are borates containing, when pure, by my analyses:—

	Natroborocalcite.	Cryptomorphite.	Silicoborocalcite.
Lime	14.20	15.55	28.69
Soda	7.21	5.61	none.
Water	34.49	19.72	11.84
Silica	none.	none.	15.25
Boracic acid	44.10	59.10	42.22
	100.00	100.00	100.00

The last two are entirely new minerals which have not yet been found elsewhere; the first I described in 1857, it is identical with the mineral called tiza, in Peru, which is imported largely into England, and to some extent into the United States for the manufacture of borax or for immediate use in the potteries. Soon after I had made the mineral known to occur here, I received a letter from a gentleman connected with the Staffordshire Potteries, in which inquiries were made about it, and a short correspondence ensued, the nature of which will appear from a few abridged extracts from Mr. Outram's letters containing interesting information. "Stoke-upon-Trent, June, 1857. I take the liberty to ask what this mineral is as I observe it is stated to contain boracic acid to the amount of 40 per cent. As this district, the seat of the pottery trade, is by far the largest consumer of this article either in its form of an acid or in the state of borax, and as its present price makes it an exceedingly heavy article in our trade any prospect of an additional source of supply will be looked to with anxious interest." "Sept. 21st, 1857; Your mineral contains nearly the same amount of acid as a specimen of the same in my possession from South America. There has lately been a large importation of borates into this district, and more of the manufacturers have been induced to use it in this state, so that although in the state of borax it is more generally used, yet it can now be pretty readily sold in the state of borate of lime. Of course it is not so valuable in this latter state and the current price in this market has lately been such that it should be delivered at Liverpool, free of charges, at something near £20 a ton. May I ask you to send me about an ounce by post, to make such a trial of it as will enable me to judge if it be suitable for the purposes of potters." The discoveries of borax in California since this letter was written must have materially altered the prices formerly obtained if it is the case that the company working there can, as they profess to be able to do, "place borax in London cheaper than it can be made there, which, at the lowest estimate, is five cents per pound." (J. Ross Browne's Report on Resources of States west of Rocky Mountains, 1866, p. 187). However this may be, the borate found here is itself valuable as a glaze, as appears from the next letter of Mr. Outram's, received soon after I had sent him a specimen. "November, 1857. I have put a portion of your mineral through the tests usually employed in our local manufacture, and I have the pleasure to enclose you a small bit of pitcher to which the borate has been applied as a glaze; as you will see, the result is really very good: the borate was applied alone, and simply passed through the potter's oven in the usual way—of course, the glazes in ordinary use, being composed of various other ingredients, possess more evenness and opacity but the fact that the Nova Scotian borate will of itself produce such a glaze speaks strongly in favour of its quality. In short it is as good as any I have seen of the same mineral." The mineral as taken from Peru is found to be an excellent flux for metallurgic purposes, and has been employed with success in the porcelain manufactory of Sevres. A good glaze has been made there by melting together one part of the borate, two of sand, and four of red lead. In fact the mineral wherever found is essentially the same and appears to be capable of effectually replacing borax in various applications while borax and boracic acid themselves can readily be obtained from it. The same is true of cryptomorphite, which has the same constituents. As regards the third mineral, silicoborocalcite, (only discovered last year), it would also no doubt form a good glaze and indeed might prove exceedingly useful for a special purpose. It appears that in glazing wrought and cast iron vessels with enamel two compositions are generally employed, one having for a base silicate of lead, the other boro-silicate of soda. The latter possesses great superiority over the former for it is not attacked by vinegar, sea-salt, or the greater number of acid or saline solutions even when concentrated, and it resists the action of the agents employed in cooking or chemical operations, **while lead** glaze gives lead to vinegar and common salt, and is darkened by cabbage, fish and stale eggs. The new mineral would probably be found to answer for this purpose just as it is found; it melts readily in the flame of a lamp to a colourless glass. I have found it in soft plaster from Newport, about six miles from Windsor, and more abundantly in both soft and hard plaster from Mr. Black's, Brookville, about 3 miles south of Windsor; the amount so far obtained is not large, the mineral is found in nodules imbedded in the plaster; from the fact of its being more abundant at Brookville where the beds of plaster are nearest the metamorphic rock it is probable any large quantity of it existing in the district will be found at the lowest part of the plaster beds. The natroborocalcite has been found at the Clifton quarry, close to Windsor, and at Newport, but is, like the silicated borate, most abundant at Brookville. In a cargo of plaster of about 300 tons, the first quarried at the latter place for 20 years, shipped last autumn, nearly every stone of a certain quality contained more or less of it, and it was found in other varieties of soft plaster; it does not occur in the hard. In some specimens of a few square inches in surface several lumps of the borate were present. The lumps were sometimes as large as an egg. I think it probable the mineral has also been found in a quarry on the Newport road, about 3 miles from Windsor. I have received an account from a quarryman there of a "stuff softer than plaster, about the size of eggs, coming clear out, and smelling like sulphur or the stones of a grist mill;" the difficulty I feel about this description being that of the borate is that odour is attributed to the "stuff." Cryptomorphite has only been found at Clifton—in small quantity.

Borate of magnesia is found in beds of gypsum and anhydrite at Stassfurth in Prussia, it also forms part of the rock at the Salt-mine of the same place; the quantity is probably considerable as the mineral was shewn in masses weighing 20 lb. or so at the late Paris Exhibition.

With regard to the other minerals mentioned as existing in the plaster-beds, glauber salt is frequently found; it goes by the name of "salts" among the quarrymen, it sometimes occurs in beautiful crystals. Common salt I have seen only on one occasion, it was in a transparent crystalline crust, it is probably often met with, but the amount is, so far as I know, not at all large: it is abundant however in the brine springs which are about to be described.

Salt from the Brine Springs of the Gypsiferous Districts. It has been long known that brine springs exist in many parts of the province where the deposits of plaster abound, but it was not till some few years ago that any attempt was made to turn them to useful account on a large scale. Salt has been made at two or three places and there are prospects of an extension of the industry on a considerable scale.

In a paper read before the N. S. Institute, in March, 1866, I gave the results of my analysis of some of the brines in question and what facts I had then obtained as to localities of others; from this paper and from information since received the following account is drawn up.

Brine Springs of Hants County. The water of a spring flowing near the gold diggings at Renfrew I found to contain 14.39 grains of solid matter to the gallon consisting chiefly of salt, there was a small proportion of earthy salts. On the west bank of the Petite river at Walton a spring issues a short distance from the bridge which has always a considerable flow of clear water, it has never been known to freeze; on a warm day in winter, the air being at 46° Fah., the water was at 44°. I found the water to contain in the imperial gallon of 70,000 grains :—

	Grains.
Carbonate of lime	14.73
Carbonate of magnesia (*small*)	undet.
Carbonate of iron	traces
Phosphoric acid, decided	traces
Chloride of magnesium	4.48
Sulphate of lime	161.16
Common salt	787.11
	967.48

the amount of salt here is equal to 1.1 per cent.

Brine Springs of Pictou County. At Salt Springs on the West river is a spring from which salt was made some 20 years ago; there are several other small springs. The salt water oozes out in many places and salt is deposited by evaporation. In water kindly procured by the Rev. A. M'Kay I found in the imperial gallon :—

	Grains.
Carbonate of lime	3.775
Carbonate of magnesia	2.932
Carbonate of iron	.181
Silica	.560
Sulphate of lime	154.730
Chloride of magnesium	27.330
Chloride of calcium	51.910
Phosphoric acid, boracic acid, Bromine, and organic matter	undetermined.
Common salt	4133.500
	4374.917

Specific gravity at 53° Fah. 1046.69

The amount of salt found is equal to 5.9 per cent. or about a bushel to the hundred gallons. The water is believed to cure rheumatism on external application.

Sutherland's River. A brine spring issues in the bed of this river its outlet being a little above the falls; it can only be got at in the dry season, it is much resorted to, its waters being drunk for a variety of diseases: it was discovered by persons observing cattle drinking at the spot.

Brine Springs of Cumberland County. A spring issuing at River Philip has been used in the manufacture of salt of which a very good specimen, dry, and of good taste and colour, was sent to the London Exhibition of 1862.

Springhill. "*Fine Salt.* Messrs. W. A. Marsters and Co., of North Wharf, have on exhibition some bags of salt manufactured at Springhill, Cumberland Co., N. S., of a remarkably fine quality, very white and pure. It resembles Ashton's fine store salt, but is probably a better article than any imported being impregnated with nitre. We understand that the salt-works at Springhill are favourably situated for manufacture, that the supply of brine is inexhaustible, and that the manufacture now in its infancy is likely to prove attractive to capitalists, the margin of profit being good, while the demand for the article is practically unlimited." St. John Journal, 1867.

Brine Springs of Antigonish County. Salt was formerly made from the Salt Pond near the town of Antigonish, where a bathing house was also established. Quite recently extensive boring operations have been carried on here which have resulted in the finding of an abundant supply of brine which is to be manufactured into salt by the Nova Scotia Salt Works and Exploration Company. Mr. Josiah Deacon, the manager of the Works, wrote to me in Jan. 1866 for information respecting brine springs. I replied, and sometime after received a letter which contains the following interesting information on the subject of the boring operations he had conducted. "Antigonish, May 22nd, 1867. You were so obliging as to furnish me with some information relative to brine springs; and as I presume you take some interest in the question as to the probability of making them available I give you the details of our last boring here. Our first difficulty and one which caused much delay and trouble was to drive by force a cast iron pipe of 9-inch bore through a bed of gravel 16 feet deep which was full of weak surface brine from the upper country. We then came to marl, red, blue, and brown, interspersed by thin bands of fibrous gypsum, but not one marine fossil of any kind. We had three boulders of magnesian limestone to cut through and nothing else to a depth of 122 feet: we did not find one drop of water. On the 12th Dec. we began to bore, for which I had fitted up a shed and a boring tower 36 feet high, so that with the aid of a stove we could comfortably work all winter. Having lost my foreman I examined every auger full brought up. The marl was so hard we were obliged to drill it with chisels and then to bring it up with the auger, but it was so dry that it would not hold in our $8\frac{1}{4}$-in. auger. I found the use of snow thrown into the bore-hole was the only means of keeping this dry marl in the auger, for water was not to be had. We then penetrated a stratum of magnesian limestone 1 ft. 2 in. and then struck a bed of gypsum into which we have now penetrated 18 **feet**, having bored in all 159 feet 4 inches. On the 27th last we had a flow of brine from a mere cleft in the gypsum; and now brine stands only 7 ft. 5 in. from the surface which we can only lower a few feet by pumping. I think I have only tapped a small vein of the great flow which I hope yet to find; and am now boring through the bed of gypsum. The brine is very pure and limpid: salt made from it is said to be of superior quality: I send you a sample. We made wooden pipes of 4¾ inches bore, and sunk them to the rock." The company above named was incorporated in May 1866. An announcement was made in June that an abundant supply of brine had been found, yielding 14¼ per cent of salt, which would become stronger, as was the case at Syracuse, New York State, where the yield was then 14 per cent. The fuel to be used was coal from Pictou county: salt was to be sold at first at $3 a ton, or 75 cents per hhd., and later at 50 cents. A profit of 40 per cent. was to be realized when working only on a small scale. The English system of working by large and shallow evaporating pans is adopted, by which, with but few hands, more than ten tons of salt a day will be made. There were in January last eight or nine buildings being erected, three of which were finished, and all the necessary machinery was being fitted up. Operations were to be commenced with vigour in the spring, and employment would be given to a large number of men.

Brine Springs of Cape Breton. About 12 miles from Baddeck, Mr. Gesner tells me, a very strong brine, giving about one bushel of salt to the hundred gallons, issues on the north side of St. Patrick's Channel.

Judique. Mr. Barnes informs me that here, where the coarse sandstones, shales, and other rocks below the plaster come to surface, several springs issue which are strongly impregnated with salt, and from one of these salt was made by the early Scotch settlers. It is not now used.

Whycogomah. From the same source I learn that at Salt Mountain there are four springs issuing from the conglomerate, all of which are strongly impregnated with salt. That on the lowest bench of the mountain is the strongest, it issues in a considerable stream, incrusting the ground in summer with a heavy deposit, having a very strong saline flavour. At the time of observation the spring was not at its full strength owing to heavy falls of rain in the autumn. Salt has been made here as I understand from Dr. Honeyman.

It is proper to call attention to the bromine which will doubtless be found on examination in all these brines. I have shewn its existence in the water of Salt Springs as indicated in the analysis given. I did not ascertain the amount as this is only got by a very troublesome operation but for so valuable a substance it was notable. In the enormous quantities of mother liquors which result in the manufacture of salt on any considerable scale it is sometimes well worth while to separate the bromine perhaps invariably present. In fact the market is supplied chiefly, if not exclusively, from this source.

In the United States bromine was first found in the bittern of the salt-works at Salina. It was afterwards detected in the salt-wells near Freeport, Pa., where it has been manufactured on a large scale for many years. The bittern here affords on an average nine drachms of bromine to the gallon, equal to about 2½ per cent. by weight; the yield varies very much at different wells. The bittern of the salt works at Schoenbeck, in Germany, contains only $\frac{7}{10}$ of a part of bromine to the 1000. It is subjected to many evaporations by which it is reduced in bulk and purified so as to contain little or nothing but the chloride and bromide of magnesium. The last process to which it is subjected sets the bromine free, and 84 lb. of the concentrated bittern yield 4 lb. of bromine at one operation. The price of bromine in London at the close of last year was about 9 shillings sterling per pound wholesale.

SALT IMPORTED INTO NOVA SCOTIA IN THE YEARS ENDING 30th SEPTEMBER.

	Quantity.	Quantity.	Value.	Duty.
1864	1,022,969 bushels	....	147,569 dollars	Free.
1865	1,004,333 ,,	1059 packages	334,134 ,,	,,
1866	1,086,735 ,,	....	189,458 ,,	,,

The whole of this salt was entered for home consumption.

Magnesia Alum. In the spring of 1862 I received from Dr. Weeks, of Brooklyn, a mineral which had been found at Parker's Mills, on the Meander River, about 15 miles from Windsor, Hants county; I found it to contain :—

$$
\begin{aligned}
&\text{Oxide of copper.} \ldots\ldots\ldots\ldots\ldots\ldots\ 0.02\\
&\text{Protoxide of iron.} \ldots\ldots\ldots\ldots\ldots\ 0.13\\
&\text{Oxide of cobalt.} \ldots\ldots\ldots\ldots\ldots\ldots\ 0.06\\
&\text{Oxide of nickel.} \ldots\ldots\ldots\ldots\ldots\ldots\ 0.14\\
&\text{Oxide of manganese.} \ldots\ldots\ldots\ldots\ 0.45\\
&\text{Potash.} \ldots\ldots\ldots\ldots\ldots\ldots\ldots\ldots\ldots 0.23\\
&\text{Slate.} \ldots\ldots\ldots\ldots\ldots\ldots\ldots\ldots\ldots\ldots 0.72\\
&\text{Alumina.} \ldots\ldots\ldots\ldots\ldots\ldots\ldots\ldots 10.64\\
&\text{Magnesia.} \ldots\ldots\ldots\ldots\ldots\ldots\ldots\ldots\ 4.79\\
&\text{Water.} \ldots\ldots\ldots\ldots\ldots\ldots\ldots\ldots\ldots 46.06\\
&\text{Sulphuric acid.} \ldots\ldots\ldots\ldots\ldots\ldots 36.33\\
&\phantom{\text{Sulphuric acid.} \ldots\ldots\ldots\ldots\ldots\ldots} \overline{99.75}
\end{aligned}
$$

MAGNESIA ALUM.

The essential constituents are the last four which are those of common alum, except that magnesia is present in place of potash. The mineral has been used in the neighbourhood of the mills in some domestic process of dyeing, and might no doubt be employed instead of common alum in various other applications: it could readily be made from the rock which contains it. This I found to be a shale or slate apparently some 60 feet thick, on the side of the river, to which it showed a nearly perpendicular face. Some of the rock was of a rusty colour, compact, and tolerably hard, this contained little or no alum; some of a blueish black colour, which readily crumbled between the fingers, showed plenty of it, partly apparently disseminated throughout and partly as a rather dense yellowish crust on the edge: the crust was in places half an inch thick, there were other more open and crystaline masses made up of perfectly white silky needles. The mineral had no doubt been formed by the action of air and moisture on the rock, and there may be deposits of it left by the evaporation of rainwater which had held it in solution; if I remember rightly the rocks dip away from the river. In the making of a large quantity of alum even the small percentage of cobalt and nickel shewn above, if constant and accumulating as a residue or by-product, would be very valuable as an additional source of remuneration.

CHAPTER IX.

LIMESTONES—MARBLES—BARYTES—MOULDING SAND—CLAYS.

Limestones. These, with the marbles, were treated of in a paper of mine read before the N. S. Institute, and published in its Transactions in 1866. In giving an abstract of this paper the London *Mining Journal* said : "The desire to turn every particle of mineral to profit becomes greater year by year, yet limestones and marbles have hitherto received considerably less attention than they are entitled to." The province contains perfectly inexhaustible quantities of limestones, presenting a great variety of qualities, of which but very few have been fully examined. These are found for the most part in the same districts as the gypsum, in the lower carboniferous beds which consist largely of them and measure 6000 feet in thickness. This system is developed almost exclusively to the N. and N. E. of Halifax, in which part of the province Dr. Dawson marks on his map upwards of 80 beds of limestone : there are only two small patches of carboniferous rocks along the whole western and southern shore : one of these contains curious and valuable limestones. In the metamorphic districts crystalline limestones are found often converted into marble of which many varieties are known.

Since the province abounds in freestones and granites which have proved excellent for house-building purposes the limestones have not so far been often applied in the same way : some have been employed in railway construction; as that at Wickwire's, near Enfield Station, where, on the side of the Truro railway, the stone was quarried to a considerable extent during the building of the extension to Pictou and used in making culverts and bridges. Limestone from Kennetcook, Hants Co., has been used at Windsor, in the foundation of the New Library of King's College which is entirely of that rock. Limestone is quarried for building purposes at several places in Antigonish county between Marshy Hope and Morristown. The economic value of the limestones will probably be always chiefly found in the making of lime for washes, mortar, cement, and agricultural purposes, and as fluxes in iron smelting for which a considerable quantity is employed at Londonderry. As regards the use in agriculture, a large portion of the best farming districts of the province lies in the formation affording limestone and except for special purposes lime will not be required in their cultivation, but it must find profitable application in such parts as are deficient in its rocks. Until recent years in which stone bridges have been made on the railways and wooden buildings have to a great extent been replaced in Halifax by those of brick and stone there could have been little demand for lime which must have been used chiefly for building foundations and chimneys because the walls and ceilings are almost everywhere made from gypsum-plaster.

On comparing the census returns of 1851 and 1861 we find of course that with the progress of the country there is increased use of lime. (In the former year there were burned in the province 28,603 casks; taking four bushels to the cask, the amount will be as follows):—

Lime burned in Nova Scotia in 1851 ; 114,412 bushels.
,, ,, ,, ,, ,, 1861; 136,848

No doubt the next census will give a greatly increased quantity. The details of the last returns are interesting, showing that five counties only burned no lime, and that the rest of the eighteen gave very different quantities, it must be stated that the Halifax return is probably to a great extent from foreign limestone and the rest from rock brought from Lunenburg county; Yarmouth also probably burns only imported rock.

Census return of lime burned in Nova Scotia in 1860:—

Counties.	Bushels.
Colchester	4,860
King's	0
Cumberland	10,635
Annapolis	0
Pictou	35,990
Hants	17,474

Counties.	Bushels.
Sydney (Antigonish)	3,232
Inverness	6,486
Halifax	26,050
Lunenburg	3,100
Yarmouth	3,500
Digby	0
Guysborough	320
Victoria	4,730
Queens	0
Shelburne	0
Richmond	406
Cape Breton	10,092
	136,848

It is calculated that during the years 1860–1866 about 10,000 barrels of lime were used per annum in Halifax in the construction of the numerous new buildings of brick and stone erected in place of those of wood destroyed by fire; a large quantity is still demanded. Notwithstanding the vast profusion of limestone in the province a good deal of that used is imported, unburnt from the West Indies, and burnt from the United States and New Brunswick. It is not easy to get at the exact amount as the returns are made under the head of "Unmanufactured Stones including Lime" in tons, casks, packages, and by number; taking, however, all the entries as "packages" to represent casks of lime averaging four bushels, and "tons" to consist of limestone, we have the following statistics:—

LIMESTONE AND LIME IMPORTED INTO NOVA SCOTIA,
In the years ending 30th September.

	Limestone.	Lime.
1864	251 tons.	68,508 bushels.
1865	8 ,,	65,404 ,,
1866	182 ,,	133,084 ,,

The price of lime varies at different ports, but that of New Brunswick averages about a dollar a barrel of 4 to 5 bushels, that from the United States about 70 cents a package, probably a barrel of about $3\frac{1}{2}$ to 4 bushels. There is no doubt that the native rocks yield excellent lime for building purposes and in some cases indeed theirs has been preferred to that from New Brunswick, while, curiously enough, the latter has in one instance been used in a locality abounding in rock affording excellent lime which could have been got at less cost. Thus lime was offered at the kiln at 85 cents a barrel, with a deduction on large demand; while New Brunswick lime was said to cost more: for some reason, however, the latter was preferred in building the new Library of King's College, at Windsor, within a short distance of the rocks whence the lime offered would have come. On the other hand, in the construction of the Railway bridges between Windsor and Halifax (one of which is 80 feet high) lime from the former place was used and found to give great satisfaction to the Engineer who pronounced it a very strong lime.* A limestone found at Indian Point, Chester, Lunenburg Co., of a deep blue colour, yields a lime which becomes as hard and lasting as cement; it is much valued in Halifax for building up the arches of kilns, a situation in which poor rock crumbles away while this remains quite hard. The lime from this was preferred to that from New Brunswick in building the extensive Wellington Barracks in Halifax, by Mr. Peters, who informed me that it is the only one yet found to his knowledge fit to use in making concrete. On Chester Basin, about 7 miles to the N. W. of Indian Point, another fine limestone is found which affords strong lime and cement, also, by its decay, umber, as mentioned under the head of paints. At Frail's, about 3 miles east of Chester, is a limestone which in some parts shews umber; it is quarried; it gives a brown lime, which sells at the wharf at two shillings per barrel.

A black limestone found at St. Peter's, Cape Breton, is said to give excellent lime, and lime is largely made from the limestone of East River, Pictou Co., and exported: it is highly valued at the neighbouring town of New Glasgow.

An important fact is mentioned in Mr. Poole's report on the Gold Fields, 1862, viz: that at two places in the northern slate and quartzite district of Queen's county, Bryden's and McLeod's, six miles distant from each other N. and S., boulders of shelly limestone were found. The limestone being easily broken could not have travelled far: search for the solid rock was recommended as in that part of the country it would be of great value for agricultural and building purposes as well as interesting from a geological point of view.

* Portland cement is being used in the Windsor and Annapolis Railway bridge over the river Avon.

Limestones as Cement Stones. Some limestones which contain a limited proportion of certain foreign minerals, as clay, magnesia, and silica, have properties not found in pure limestones. After these are burned at a proper temperature they either do not slake at all or very slightly and slowly; when afterwards moistened, but especially if pulverized, they absorb water without becoming hot or swelling up and after a length of time varying from hours to days they become hard, and when this takes place under water they are called hydraulic cements.

English cement stones are stated usually to consist of:—

```
       Carbonate of lime.................65.70
       Protoxide of iron................. 6.00
       Silica............................18.00
       Alumina........................... 6.60
                                         ——————
                                          96.30
```

or, generally, to contain from 8 to 25 per cent. of the foreign substances named. (Robert Hunt.)

According to Deville * a dolomite rich in carbonate of magnesia (the rest being carbonate of lime) when calcined below a dull red-heat, powdered and made into a paste, forms under water a stone of extraordinary hardness. Some stones exposed to the action of the sea remained unaltered. Dr. Calvert about the same time gave † the following analyses of rocks employed by the Great Dinorben Mining and Cement Company, in Anglesea, where some magnesian limestones are worked which make excellent hydraulic cement and stucco. The results were:—

	Hydraulic cement.	Hydraulic lime.	Stucco.
Carbonate of magnesia	61.15	55.23	15.85
Carbonate of lime	21.41	33.99	72.23
Carbonate of iron	8.76	3.85	3.21
Silica	5.58	1.26 ? }	2.70
Alumina	2.07	2.27	
Organic matter and water	1.10	3.40	6.00
	100.07	100.00	100.00

The analyses shew that the hydraulicity of the rocks is in proportion to the amount of carbonate of magnesia they contain. The

---
* Chemical News, XII., 287.
† Chemical News, XIII., 6, where the amount of silica in the second analysis is given as 5.58. I have ventured to alter this as there is evidently an error, and, from the remark of M. Deville given after, it is probably in the silica.

rock best for each purpose contains the amount of this ingredient stated above. The author states that he has compared the strength of these products with the best Portland cement and blue lias limestone and found them quite equal, though very different in chemical composition. Dr. Calvert supports M. Deville in the statement that the calcination must be effected carefully. The heat must be raised to redness gradually and kept there till all the carbonic acid is driven off, a higher temperature destroys the hydraulicity. The calcined product must be ground very fine: the finer the powder the better the cement sets. M. Deville was of opinion that while the magnesia in the hydraulic cement above was the main cause of the hydraulicity the silica present must also be beneficial.

In view of the preceding statements it is important to see what is known of the composition and qualities of the various limestones met with in the province. The fact is an exceedingly small number of them have been analysed, even qualitatively, but several have been reported to possess hydraulic properties and some are known to make excellent cements. The limestones of Walton and Teny Cape, Hants county, found holding, or in the neighbourhood of, manganese often contain magnesia but in what quantity is not known; their hydraulicity has not been tested. I have been told that some of the limestones of Windsor afford hydraulic lime, and the same is said of some from St. Peter's, Cape Breton. The "paint stone" of Chester Basin, an impure limestone mentioned at another page as weathering to umber, consists of carbonates of lime, protoxide of iron, manganese, and magnesia, with other ingredients among which are pyrites, sand, and apparently bitumen; it is dark blue, nearly black, when first extracted, and when carefully burned becomes dark brown in parts. In one specimen I tried the burnt mass quickly absorbed a small amount of water, grew warm, but did not fall to powder; when broken it shewed some white portions; on grinding, a uniform brown powder was easily obtained which when mixed with water to a stiff paste became solid in half an hour and in one hour or so a cake of it was dropped from a height of five or six feet and only a small fragment was detached. The next day the cake was thrown to the ceiling and only broke on falling a fourth time to the floor. The cement appeared to get harder on exposure in the air. The addition of sand gave me only a crumbling mass. I understand that by proper treatment hydraulic cement has been made from some of the rocks of this locality which are found at different points for some miles in an east and west direction along the Basin.

Some of the limestones on the property of the Onslow East Mountain Lime and Manganese Company contain foreign ingredients and may be found to be hydraulic. The chief impurity appears to be clay, there is a small quantity of magnesia in some of them, some are bituminous. Mr. Barnes, who as well as myself reported on this property, says the limestone is very similar to the Canadian limestone affording strongly hydraulic lime, and, that while the water cements used here are imported at an expense of $4 or $5 a barrel, if the lime found here should prove a good cement, it could be sold at from 50 to 75 cents a barrel.

The soft blue limestone from Kennetcook, Hants Co., used in building the foundation of the new library of King's College, I find to contain a considerable amount of magnesia, with some protoxide of iron and some siliceous matter. A limestone from Arisaig, Antigonish county, examined in connection with Dr. Honeyman's geological survey, afforded me :—

Carbonate of lime......................74.64
Carbonate of magnesia.................. 4.84
Oxide of iron, with a little alumina....... 5.05
Water 0.16
Clay and sand........................13.82

98.51

and it was stated in my report that the amount and nature of the ingredients other than carbonate of lime are in favour of a certain amount of hydraulicity.

Mr. G. Lang informed me that Shubenacadie affords a limestone from which lime was used twelve years previously in building a chimney for a steam-engine, and that after the lapse of that time the work under water could not be separated. He believed that the lime takes the first place on this continent for masonry and all exterior work. It slacks with unusually little water, and takes as much sand again as any other used in the country, making a mortar which is better than any cement except the Portland and resists the severe frosts and sudden thaws much better than that made with lime from St. John or West Indian limestone. Mr. Handley, of Halifax, shewed me a cement he had used in putting together firebricks, which he had made from a rock found near St. Peter's, Cape Breton, by careful burning, grinding, and mixing with sand: he found it a very strong cement. During the construction of railways and other public works in (old) Canada one manufacturer made on the average 8000 bushels of cement annually.

Limestones for Manure. While magnesia is found to be so valuable a constituent in limestones to be used in making hydraulic cement, its presence is considered prejudicial in lime for manuring purposes, and I have heard the limestones of Windsor objected to on this account. Some of the deposits in this neighbourhood are certainly nearly pure carbonate of lime. I analyzed a specimen from the fossiliferous bluff on the Avon, on the property of Otis King, Esq., and found it to contain:—

Carbonate of lime......................97.64
Carbonate of magnesia 1.10
Oxide of iron.......................... 0.07
Phosphoric acid .... trace.
Clay, sand, and silica................. 0.68
                                        ——
                                        99.49

For the sake of comparison I may state that in Prof. Anderson's recent work on Agricultural Chemistry, the analyses of two common limestones are given as examples of the composition of these rocks and 1.61 and 7.45 are the respective per centages of carbonate of magnesia. There are many limestones in the province as at other parts of the banks of the Avon near Windsor, at Stewiacke, and at Brookfield and Gay's river, Colchester county, which closely resemble that above analyzed in being made up almost exclusively of marine shells; they are all probably equally pure.

The presence of phosphoric acid in some of the limestones at different localities, occasionally apparently in notable quantity, may add much to their value for agricultural purposes as Dr. Dawson mentions with regard to some found in the coal measures of Joggins, Cumberland Co. I examined a specimen which I got from a bed on the beach there and found a very decided amount of phosphoric acid, it was from a dark coloured bituminous rock. These would, as stated in Acadian Geology, be worth about three times as much as ordinary limestone, and the riches* of the beds might possibly be sufficiently appreciated on trial to allow them to be profitably worked. I found a small quantity of phosphoric acid in some of the bituminous limestone of Onslow, and a very small quantity in some not bituminous.

Limestone as a Flux. At Londonderry the amount of limestone used was in 1861 given as 200 bushels required to smelt every ton of iron, of which there were at that time about 1,200 tons annually produced. When the Nictau works were in operation limestone was imported from New Brunswick and then conveyed by land carriage a distance of some eleven miles over a mountain road. In the event of the works being reopened railway transportation of the limestones of the carboniferous districts at the east end of the Windsor and Annapolis line may be found of great advantage.

Marbles. These have long been known to exist in various localities but none of them have been actually worked; attempts have been made to bring two of them into use but the results have not been satisfactory, from the fact that the specimens obtained, while very beautiful, have not proved to be free from flaws. This may be due to the circumstance of the rocks being found in metamorphic and disturbed regions; also, however, as none of the marble deposits have been worked in depth, to the actions which are found to influence all rocks near the surface. On this latter point Prof. Ansted says : "All building materials of the nature of stone, forming part of the earth's crust as a mass of rock, if they come to the surface at all, are invariably injured and altered near the surface." In the case of marble, which is not absorbent of water on account of its close texture, the flaws might be caused by the expansive action of frost on water in crevices.

The marble best known here is that found in the metamorphic rocks of Five Islands, Colchester Co., near the Basin of Minas. It is situated on the side of a steep mountain at the base of which runs a brook in whose bed are many fragments of marbles of at least two kinds. The marble found in largest quantity is pure white in colour, of excellent grain, surpassing in beauty when polished, according to Messrs. Wesley and Sanford, marble-workers, of Halifax, the Italian marble. About 1852 a gentleman was sent from England with two quarrymen to extract a block.

The party remained some months and finally a block of considerable size was shipped, at an expense, it is stated, of about £1000. The explorer is reported to have stated that the marble was superior to any he had seen from Carrara, but on the arrival of the block in England it was pronounced to be unserviceable from being shattered. The condition is considered here to have been brought about, in part at all events, by the block having been blasted out. I have been told by a resident in the neighbourhood at the time of the operations that a good deal more might have been done at the same expense, and, on the whole, the rock can hardly be said to have had a fair trial. Even if better conducted operations do not show that large masses of good quality can be got, at least it is probable that smaller blocks suitable for busts and like sized objects may be obtained. The other marble found here is **green and white**; it is not very well thought of by marble-workers.

The second trial made on marble was in Cape Breton, in the neighbourhood of Bras d'Or Lake. Here the rock is found on the top of a mountain for a distance of about six miles. Samples were taken out about three years ago, polished, and sent to New York, they were pronounced shattered, and operations were given up. It is curious, however, that, as Mr. Lordley, who was one of the company concerned, tells me, the boulders found in the district are solid. The colour was white with red stains, a green stained white variety is also found in the same region. One great advantage of the locality is that a gentle descent of two and a half miles leads to a shipping place.

The best display of the marbles of the province was made at the **London Exhibition of 1862**, when thirteen specimens were shewn from eleven localities. Parrsboro, in Cumberland county, furnished a purple coloured marble with green spots of serpentine; Cheverie in Hants county, a red banded variety; Five Islands, the two kinds described above; Onslow, Colchester county, a red and white mottled, and a chocolate variety; Pictou county a greenish coloured from East River, and a grey patterned variety from Fraser Mountain, near East River; Cape Breton sent a white with black veins from Whycocomah, a red and a clouded grey from Craignish, a white and green from George's River, and a black marble from some other locality unnamed. The grey marble from Fraser's Mountain is concretionary, it was shewn in a polished specimen, of about a square foot in surface, (due to the firm of Wesley and Sanford, of Halifax, before mentioned, who very liberally polished various other specimens for the Exhibition Commissioners) it exhibited concentric waved bands in separate sets whose outlines somewhat resembled expanded flowers. So far as I know this marble is unique, and a specimen I have, a present from the firm named, is an object of great admiration to mineralogical visitors; it would make fine inlaid work. The quantity of the material, I understand, is considerable but it is not all equally beautiful when polished.

Barytes.—This mineral, called also heavy spar, is found in several localities and of various degrees of purity. The principal uses to which it is put are the making of porcelain, the adulteration of white lead, the making of permanent white pigment, and the enamelling articles made of paper, as cards and collars, and giving a peculiar surface to room-paper. A statement in the Quarterly Journal of Science, of January 1867, gives 40 tons a day as the consumption at one factory in New York for enamelling collars. A dealer informed me that he could sell 10,000 tons a year in that city, and gave the value as $20 a ton crude, and $35 a ton refined mineral. From the same source I learned that 50 tons a day are used in Connecticut, where it is abundant. The consumption was very great in England in 1866, when the selling price of the rough delivered was from 25 to 30 shillings per ton, that of the ground and bleached probably 70 shillings sterling. The only mineral that sells well is that which is quite white and altogether or very nearly free from ores of metals such as iron and copper, which impart colour when it is ground and put to use, and the "bleaching" consists in the removal of such impurities. It is useful to know that the mineral with reference to which the foregoing prices were given in England was stated to have this composition:—

Sulphate of barytes..................82.46
Silica, sulphate of lime, etc..........17.54
100.00

a note being added that the mineral was very free from iron, and that some portions of it would not require bleaching. The amount of barytes raised in the United Kingdom in 1861 was 11,451 tons.

Barytes of Five Islands, Colchester County. The mineral occurs here in numerous irregular veins or pockets in the slates of East or Bass Rivers. It is sometimes found in very beautiful crystalline masses, one of these, weighing perhaps 150 lb., was sent to the Paris Exhibition. It is associated with calc spar and carries in parts a little specular iron and copper pyrites. It has been worked to some extent. I visited the mine, which is two or three miles from the high road, in the summer of 1866 and found it in operation. The veins from which the mineral was being taken were nearly vertical, one had a thickness of a few inches and a good deal of barytes was being taken out, another, in the opposite side of a high hill, from which much had been removed was about three feet wide in parts, I learned there was another exposure from which a large quantity had been got. The mineral was extracted from the face of the rock by simple quarrying. About 30 tons had been got out in three weeks; some of the mineral was quite white, some had a red tinge, and some contained specks of copper pyrites. It is estimated that about 500 tons in all have been exported to the United States. Since my visit fresh discoveries have been made, three new veins having been found above the old ones. The mineral, of which I have seen a specimen of excellent quality, is said to be much superior to that formerly obtained. The Colchester Baryta Company has issued a prospectus containing favourable reports by Professor Wyckoff and others.

Barytes of Stewiacke, Colchester County.—On the banks of the Stewiacke river, about 4 miles from the Brookfield railway station three veins of barytes are exposed on the surface, in a country rock of red sandstone, having an average thickness of 18 inches. The mineral has been got out in some quantity, the largest amount having been removed several years ago. A shaft of about 40 feet was sunk on the first workings by quarrying and one vein was found to thicken very considerably in depth. Last summer the shaft was emptied of water and a few tons of mineral raised the greater part of which is said to have been perfectly white. A specimen of some pounds weight shewn to me, reported to have been taken from the surface, was white throughout, or with a greyish tinge in part, but perfectly free from pyrites and other metallic minerals; this latter character is said to belong to the whole deposit. Mr. R. G. Fraser, to whom I am indebted for these details, says he has no doubt of the mineral being abundant and he adds that there is a mill within 500 yards of the shaft at which the mineral could be ground; the road to the mill is on an incline from the shaft. It is estimated that 1200 tons, in all, have been taken out.

Barytes of Brookfield, Colchester County. On the mining property controlled by Mr. C. Annand, in the neighbourhood of the large boulders of iron ore described in a former page, is a deposit of barytes, probably of some 15 feet in thickness, exposed on the side of the hill. It is mixed with iron ore and most of it as seen has a reddish colour but one specimen I obtained was quite white.

Barytes of other localities. I have specimens of impure mineral from the bank of the west arm of the River Avon, Hants county; from the peninsula of Halifax; from the Onslow East Mountain Company's property where I found it in a detached mass; and from Frenchman's Brook, Arisaig. This last was among the minerals examined in connection with Dr. Honeyman's survey, it was mixed with red oxide of iron; the broken mineral showed a red colour partly in the mass and partly in layers on the surface of fracture, and, when ground, gave a white powder with a decided tinge of pink; the colour was partly removed by acid. These last named being all surface specimens can hardly be taken as shewing the true nature of the deposits. Mr. Barnes informs me that about 14 miles north east of Cheticamp, Cape Breton, a vein of barytes more than 18 inches thick occurs in a highly contorted black slate. Huge masses of the mineral containing an admixture of small particles of slate are lying on the beach. Barytes is found, perhaps invariably, though not in quantity to be worth working, with the manganese ore of Hants and Colchester.

Moulding Sand. This is a material of the first importance in metal casting, a good quality of it being eagerly sought for and sometimes obtained from distant places, of course at considerable expense. What is required at the foundry is a sand having sufficient cohesive power when moist to remain firm in the flask and this quality is found not in a pure siliceous powder but in a

K

sand containing an admixture of foreign substances such as clay and oxide of iron.

It has been thought that much of the beauty of the famous Berlin iron-castings, which are unrivalled for delicacy and exquisite finish, depends upon the sand employed in forming the moulds. Hence the sand collected about an ornamental casting shewing the perfection to which the processes have been brought was analyzed at the Royal School of Mines, in London, with these results :—

>   Silica............................79.02
>   Alumina...........................13.72
>   Protoxide of iron.................. 2.40
>   Lime.............................. trace
>   Magnesia.......................... 0.71
>   Potash............................ 4.58
>   ───────
>   100.43

it was probably a decomposed granite.

Different qualities of sand are required, respectively, in the casting of iron, of brass, and the fine kinds of brass-casting, such as tubes. For the former purpose, the foundry of Messrs. Dimock, at Windsor, uses to some extent sand from near the railway depot about half a mile distant, sand from New York being also employed. Dartmouth affords beds of sand one of which is used by the brass founders of Halifax. Promising beds of sand are found at Onslow, Colchester county, and at Chester Basin, Lunenburg county. Some time ago, the American Tube Works Company, in Boston, having been in the habit of using sand from near Birmingham, England, and from Baltimore, U. S., which last contained 4 per cent. peroxide of iron and 3 per cent. clay and cost $10 per ton, were anxious to find some sand in this province which would answer as well and cost less money. After the trial of several qualities one found in abundance on the property of Mr. Pellow, at Windsor, Hants county, was proved to give satisfaction and, in 1866, a shipment to the amount of 250 tons was made. The sand is red, it occurs in a bed some 8 feet thick, about three feet beneath the surface, overlying a thin bed of grey sandstone which rests upon the great bed of gypsum Mr. Pellow is now working in the Clifton quarry. Only 50 cents a ton were charged, but on another occasion it would be necessary to make the price 75 or 80 cents.

Bath-brick Sand.—The sand above mentioned as occurring at Chester Basin I am led to consider promising for moulding purposes from its fine grain and from its having sufficient cohesive power to admit of being made into bath-bricks. Mr. R. D. Clarke, who, as before mentioned, used to make paint from the "limestone" of this neighbourhood, pressed the sand into bricks which he dried in the sun and sold as bath-bricks with the mark of "Douglas Brick," about 1850. The sand, of a light yellow colour, is in a bed of some feet thick close to the road which passes within a short distance of the shore of Chester Basin.

Mortar-sand.—Sands for making mortar exist in many places, but the district through which the coach-road passes in Kings and Annapolis counties is particularly rich in sand which seems especially adapted for this purpose. With regard to the quality of that in the neighbourhood of Kentville, the engineers of the railway from Windsor to Annapolis have found that from its hardness and cleanness it is the best for making mortar they have ever met with.

Clays.—Large deposits of clay, suitable for making bricks and pottery, are found in many parts of the province and the numerous coal fields afford abundance of fine clay, some of which is undergoing trial for the manufacture of fire-bricks. Extensive brick yards have been in operation for some years and potteries have been added to these at which a variety of ware has been made. The census returns of 1851 and 1861 gave the following:—

Number of bricks made in the province, 1850......2,845,000.
     ,,   ,,   ,,   ,,   ,,   1860......7,659,000.

The subjoined table from the last census, 1861, gives the number and value of the bricks made in each county, and affords an index to the localities of the clay districts, as then recognised.

BRICKS MADE IN NOVA SCOTIA IN 1860.

Counties.	Number.	Value in Dollars.
Halifax	1,092,000	7,944
Colchester	760,000	4,905
Cumberland	335,000	2,465
Pictou	411,000	3,286
Sydney	180,000	1,309

Counties.	Number.	Value in Dollars.
Guysborough	150,000	1,212
Inverness	47,000	237
Richmond	0	0
Victoria	6,000	30
Cape Breton	5,000	30
Hants	1,555,000	13,590
Kings	775,000	4,818
Annapolis	880,000	4,253
Digby	148,000	919
Yarmouth	1,200,000	6,000
Shelburne	0	0
Queens	25,000	150
Lunenburg	90,000	555
Total	7,659,000	51,703

The following statement from the Trade Returns shews the number and value, when this is given apart from that of other "manufactured stone," of the bricks imported and exported in the years ending 30th September.

Year.	BRICKS IMPORTED TO N. S.		EXPORTED FROM N. S.	
	Number.	Value.	Number.	Value.
1864	1,188,685	$11,565	48,000	$451
1865	1,764,662	17,743	19,480	not given
1866	869,500	not given	740,000	455
1867 (9 months)	377,050		28,000	392

The average value of bricks appears to be ten dollars a thousand, though there are wide departures from this price upwards and downwards both in imports and exports; portion of the totals is no doubt fine brick. Of course there is not so large a demand now as there was a few years ago when the burnt houses of Halifax were being restored in brick and stone and the side walks were being paved with brick; from this cause and the importations from New Brunswick and "Canada" together with the large home production the market has lately been overstocked and a large proportion of the bricks made the last year or two remains unsold, though the price of those made along the eastern railway was reduced to $7\frac{1}{2}$ dollars the thousand delivered at Richmond.

As appears from the census returns, brick-clay beds are found in many parts of the province. Some of the most extensive occur in the eastern part of Hants county; along the line of railway to Truro, from Windsor Junction to some miles beyond the Shubenacadie station, these beds are worked by various companies at different points. At Mr. Malcom's works, situated close to the railway, both bricks and pottery are made. The brick-clay covers a large tract to an average depth of nine feet on the side of the Shubenacadie river; in 1867 about 1,300,000 bricks were made. The fire clay also found close to the railway appears to run in beds, of no great thickness, dipping at a considerable angle. Fireclay from Nine Mile River was formerly used here extensively, it was in a bed 12 feet thick, mixed with vegetable remains and resting on a quartz gravel. The fireclay now used is brought from Musquodoboit where it occurs in a bed known to be 25 feet thick, and of unascertained further depth, in bands of various colours; it is blue on the top, below are seams of white, black, and yellow clays. The distance is 18 miles by road and other 10 by rail; the clay costs the company about $3 per ton: some 500 tons were used last year in making fire-bricks (to the number of about 10,000) drain-pipes, chimney-pots, stone-tubes and other large articles. Operations commence the first week in May. Mr. Malcom sent fire-bricks, drain pipes and pottery ware to the London Exhibition of 1862. Of the bricks exhibited the report of Rev. Dr. Honeyman, agent for the Nova Scotia Commissioners, states: "All the bricks were considered as excellent and well-made, and were highly approved of both by Englishmen and foreigners."

Nash's Brick and Pottery Company have works one mile beyond the preceding. They have 8 or 10 acres covered with 10 feet of clay without a stick or a stone. About 1,000,000 house-bricks were made in 1867: they have means of turning out three times that amount. A few thousand fire-bricks were also made, and used chiefly in building their pottery ovens; 5000 only were sent to the Albion Coal Mines.

Mr. Lang has a brick yard 12 miles N. E. of the last named; he has 23 acres of clay bed 60 feet thick and of which the bottom has not been reached; he made in 1867 about 600,000 bricks. By his neighbour, Mr. Murray, there were made that year about 400,000.

Near Windsor Junction a brickyard has been established by Mr. Grove: he has a thickness of clay in one part of 25 feet, and has found no bottom. In 1867 about 300,000 bricks and a few flower-pots and chimney pots were made.

On the eastern shore of Halifax harbour is situated the Wellington Brickyard, a large establishment under the management of Mr. T. Scarfe, who has kindly furnished me an account of the works as they were before they were destroyed by fire about three years ago and as they now are: with reference to the present condition he says: "Feby. 4th 1868.—We have during the last two years been erecting new buildings and fitting up new and improved steam engine and machinery. We can now manufacture upwards of 40,000 bricks a day, and have facilities for extension should a greater demand arise. Our works are about three and a half miles from the city by water, the shore is very bold with good anchorage and with our extensive wharfage we have every facility for loading ships of any burden. The clay which is intermixed with sand, fine gravel, and large stones, (but little or no limestone) is taken from a hill running N. W. and S. E. parallel to the harbour, and apparently one of the most prominent lateral moraines which abound all along the eastern shore. We work the whole face of the hill in order to combine the various ingredients and by that means greatly improve our manufacture. The face of our workings is about 150 yards long with an average height of 10 feet, the base being 11 feet above high water level and on the same plane with the upper portion of our yard. The deposit extends for about 400 yards beyond our present face. We manufacture about 2,800 cubic yards into 1,500,000 bricks yearly. We have several other hill formations on our premises but their material is not so good as that we are now using.

"The clay having been dried is passed through large revolving screens to free it from the stones, it is ground by powerful rollers which thoroughly pulverize and incorporate the various minerals previously to their being ground with water in the pug-mills: the bricks are thus rendered homogeneous throughout. We are still burning our bricks in open kilns with soft wood which gives the most satisfactory results with our stock. This method of burning by open kilns is not practised in England but is much in vogue in Canada and the Northern States; the fuel is fed in from time to time during the burning as to the furnace of a closed kiln.

"Our kiln sheds are capable of holding 1,000,000 bricks. We

claim for our manufacture superiority in cohesion, solidity, and colour, above any others made in this province. About 40 men are employed upon the works and we could often give work to a much greater number if they could be procured when required."

At the local exhibition held in Halifax preliminary to that of London, 1862, Mr. Scarfe obtained a prize for bricks of which the following is his description. No. 1. Pressed bricks made expressly for the face of buildings and passed through a powerful combination press after the greatest amount of the shrinkage has taken place in the moulded clay, whereby a fine, smooth surface with sharp outlines is obtained, combined with increased density, rendering them more durable and less porous.

No. 2. Slop stocks, being moulded with water and struck by hand, forming a very clean and excellent brick suitable for the exterior of buildings.

No. 3. Drain bricks, moulded for 12 inch barrel drains, the beds of which radiate from the centre, whereby a great saving of cement or mortar is obtained.

No. 4. Common or sand stocks made by machinery; a thoroughly well burned brick which from the roughed surface caused by the moulding sand forms a strong and adhesive bond with mortar or plaster."

There are also made bricks of any other desired form, and paving tiles for kitchen, dairy, and cellar floors.

The favourable opinion entertained by those who saw the bricks exhibited in the Nova Scotian court of the London Exhibition, in 1862, has been before given. Mr. Scarfe received on that occasion "Honourable Mention" for "good quality of common and pressed bricks, and drain tiles."

Bricks are also made by the Industrial Manufacturing Association and by Mott & Co. in the neighborhood of Halifax.

In Pictou county, Messrs. Primrose and Rudolf have a **brickyard** near the town of Pictou from which they turn out about 500,000 bricks annually. Pressed as well as common bricks are made. In the neighborhood of New Glasgow are extensive beds of clay in the coal measures which have been opened up and found to be of excellent quality for making firebrick and pottery; according **to Mr.** Haliburton the clays in this district have been pronounced **by par-**ties in Staffordshire unsurpassed by any in England.

The Crown Coal Pottery and Brick Works Company have made considerable preparations for extensive operations about a quarter of a mile from the town of New Glasgow: among the articles now made are flower-pots, flue conductors, chimney pots, butter coolers, Parisian Medici vases, glazed and unglazed earthenware, glazed stoneware, and fire-bricks.

About two miles from Windsor, on the St. Croix, Hants Co., is found a deposit of very white clay, which is used as a wash in the neighbourhood and also at Messrs. Dimock's foundry at Windsor for stopping the crucible and lining the melting pots.

Mr. Poole mentions in his report of 1862 that he had a sample of Kaolin or pipe-clay of very fine quality and very white given him from the banks of the Sabbattee Lake, 4 miles from Chester, Lunenburg county; the water was too high to allow of exploration.

At the late Provincial Exhibition an interesting display was made by several of the companies above mentioned and well deserved awards were obtained for various articles. As the official prize list has not been published I do not venture to give details which might be incorrect.

CHAPTER X.

BUILDING STONES—STONES AND MATERIALS FOR GRINDING AND POLISHING.

It has been mentioned that limestones have not been much used as building stones; where stone has been employed in this way preference has been given to granite and freestone. Since the enactment was made which compelled the use of stone in replacing the numerous wooden buildings destroyed of late years by the frequent fires in Halifax there has been great demand for these stones in the city. Their character is well shewn in Granville Street the new part of which is justly considered very handsome. Further instances of their employment are seen in various parts of the city and harbour in private and public edifices, particularly in some of the Banks, the Province Building and new Post Office, and the fortifications belonging to the Imperial Government. In the country, except at Pictou, stone buildings are the exception Windsor, exclusive of the Collegiate Academy, and the New Library of King's College, has but one structure, a house, of stone. The freestones are much esteemed in the United States where they have been largely used. The building stones exhibited at the International Exhibition, London, 1862, received the award of Honourable Mention "for goodness of quality," and at the International Exhibition, Dublin, 1865, Mr. G. Lang, obtained Honourable Mention "for a well selected series of good building stones." In fact there is no country, probably, better furnished with the varieties of stone suited to the purposes of the civil engineer, the builder, and the architect. Those adapted for decorative work, except as regards carving, have not been employed at all.

Granite. This is found, as before mentioned, in many parts of the southern metamorphic districts of the province. It is abundant in

Shelburne county, where it occurs at Wood's and Shag Harbours, and extends inland for some distance; it is seen at Port Joli and Port Mouton, Queens Co.; about Chester, Lunenburg Co., where huge boulders with a face of some 30 or 40 feet occasionally abound near the basin, and a few miles to the east it forms a thick dyke or ridge separating Mahone Bay from Margaret's Bay and extending inwards to near the Avon and Ponhook Lakes. Boulders of it are frequent in the drift of Windsor and abound at Butler's Mountain about three miles south of the town, probably on the surface of the bed rock of the same character, and a few miles from Windsor it may be seen occasionally in numerous boulders on the railway thence to Mount Uniacke, near which place it is in situ. Two of the railway bridges in the district traversed are built of these boulders, one of these, at Big Brook, is 80 feet high. The district east of Margaret's Bay terminating at Cape Sambro, Halifax county, contains several varieties of granite, and near the North West Arm of the harbour of Halifax the government granite quarries are situated. Mr. Gossip (in a paper on the Rocks of the Vicinity, Trans. N. S. Inst., 1864) says, "there is a gap between the granite as it comes down to the arm and as it appears again eastwardly. There is not a particle of the rock *in situ* on the peninsula of Halifax, or on the Dartmouth shore, this is the case also with the coast for a considerable distance and the interior also, so that there is a large tract of country in which granite is absent. The nearest point at which it reappears to the east is probably Lake Thomas." It is seen again at Musquodoboit River and extends east as far as Great Ship Harbour Lake; beyond this it is found in Guysboro' county about St. Mary's River, near Sherbrooke, and the lakes about Indian Harbour, and a large part of the peninsula terminating at Cape Canseau is occupied by white fine-grained gneiss with veins and masses of granite.

The devonian and silurian districts, Dr. Dawson says, have not so much granite as the older metamorphic region. The most extensive appearance seems to be in Annapolis county, where the slates are interrupted by great masses of the rock which form the hills along the south side of the Annapolis River from Paradise to Bridgetown and, though not continuously, nearly as far as the town of Annapolis. This granite is hardly to be distinguished from that of the south district except that it is perhaps more felspathic and less largely and perfectly crystalline. In the Cobequid Mountains, in Cumberland and Colchester counties, masses of red, flesh-coloured and grey syenite, seen rising rapidly to the height of several hundred feet, have often been described as granite, some of the varieties are often fine grained and appear to pass into greenstone. It may be conveniently mentioned here that these rocks and the porphyries of the district are suitable for building and ornamental purposes. I have had a few specimens of the syenite and greenstone roughly polished and the results were such as to shew that the qualities of these rocks are well worth trial on a large scale.* Syenite when polished forms, according to Dr. Feuchtwanger, the most splendid of all ornamental rocks. 5,000 houses are supposed to be built of it in the city of New York, where it is often called granite; the distinction generally observed is that of the former has hornblende in place of the mica of the latter. Granite is quarried at Shelburne, and is much approved in Halifax as being very free from iron stains.

In the neighbourhood of Halifax granite is quarried at the government quarries near the North West Arm and at private quarries at Birch Cove. Mr. Gossip says : " there are two descriptions of granite within a short distance at the former locality, one is much harder and finer grained than the other, and they differ in colour. The granite nearest the blue rock and slate becomes somewhat porphyritic and acquires a light greyish blue tint; wherever it approaches the lower rocks it grows harder and, though in some places of good quality, is more difficult to work. There appears to be a well defined line of division between the two kinds of granite. A natural gully leading from the south government wharf to to the granite quarry in the hill beyond has on one side the purer description of which it is but truth to say that there is nothing superior anywhere to be found of granite rock. Blocks of the largest dimensions may be quarried here, and the supply seems inexhaustible." From the Board of Works report for 1864 it appears that the cost of rough granite at the Penitentiary, where it is cut by criminals, is $2.30 per ton, the cost of cutting is given as averaging 25 cents per superficial foot.

I visited the quarry at Birch Cove belonging to Mr. R. Davis in

* I have since been informed that four columns in front of Molson's Bank at Montreal, much admired for their beautiful appearance, are made of polished red syenite from this province.

company with that gentleman in 1861 and there also the supply of granite is very large and the quality excellent: a specimen of one foot cubic dimensions was exhibited by Mr. Davis, in London, the next year; it was cut and polished, and was very handsome.

Freestones. These are found in the greatest abundance: valuable varieties exist in many parts of the carboniferous districts of Nova Scotia proper and Cape Breton, some of which have been largely quarried for home use and for exportation.

Pictou County. Gray Freestone fit for building purposes is found in a great number of places in the coal formation and is quarried at present chiefly at two points on the Saw Mill Brook at the head of Pictou harbour, the "Acadia Quarry," and "M'Kenzie's Quarry." These quarries have yielded large quantities of stone, and a railway and loading pier three quarters of a mile in length have been constructed. Many structures have been built of the stone in the large cities in the United States; that from the M'Kenzie quarry was used in the lower part of the New Post Office in Halifax: it is cheap, durable, and of a pleasant reddish colour. The price of the stone depends much on the dimensions, for the usual size of building stones the price is about $4 a ton shipped. Previous to the late war in the United States the amount annually shipped was, Mr. Ross informs me, from 4000 to 5000 tons; the war destroyed the trade and since 1861 the Acadia quarry has scarcely been worked. Last year, 1867, the quantity taken from both quarries did not exceed 400 tons. In the seven years ending 30th September, 1866, there has been exported from Pictou, stone, chiefly building, to the value of 25,094 dollars.

Cumberland County. At River Philip quarries have been worked for three years by Messrs. M'Donald, who have shipped annually about 1,500 tons of stone, worth $5 a ton at the quarry. The stone has been used at the new Post Office, Halifax; the Bank of Yarmouth; the Volunteer Monument, Toronto; the R. C. Cathedral, Harbour Grace, Newfoundland; in P. E. Island; in Montreal; and in Portland, Maine. There is a deposit of stone of beautiful colour on this river not yet opened.

At Wallace valuable beds have been quarried for many years by

Mr. Batty. The **stone** is light in colour, easily worked, of good quality, and admits of very elaborate and delicate sculpturing. It has been largely used in Halifax; the Union Bank and parts of the new Post Office are of this material. The price is $2 a ton **at** Wallace, the amount exported is, when given in the returns, not separated from that of grindstones.

At Tatamagouche freestone is also quarried and to a small extent exported. In the seven years ending **30th** September, 1866, there have been exported from Cumberland county, exclusive **of** the Joggins, stone, chiefly for building, to the value of 21,211 dollars.

Hants County. Sandstones suitable for building purposes occur **at** Shubenacadie, Falmouth, Windsor, Kennetcook, and no doubt other places. The Kennetcook stone is not of very fine grain, but **it** is easily **worked**, is close to the river, and its colour which is yellow, and rather brown in parts, is warm and agreeable. This stone was used in building the new Library of King's College, Windsor, and the Bishop of Fredericton, whose architectural taste is well known, expressed, in my hearing, great admiration of its colour.

The Falmouth stone and that from Kennetcook are being used for the new bridge over the Avon, at Windsor, for the railway thence to Annapolis; the former is raised about **2 or 3 miles from** the river. The amount required is estimated at from 8,000 to 12,-000 tons.

Colchester County. A valuable quarry of building stone exists about 4 miles from Folly Village.

Cape Breton. Freestone of good quality is obtained **at** Port Hood **Island, Margaree,** Whycocomah and Boulardarie.

Flag Stone. A large quantity of stone has been quarried at the North West Arm, and used as a flagstone. **Mr.** Gossip says the Flag Quarry lies nearly opposite the Chain Battery, a short distance from the wharf where the stone is shipped. On the top the rock **is** a good deal broken; at a depth of 6 to 10 feet excellent building material is found in beds from 3 to 6 feet **thick, it is a** bluish grey quartzite, is often micaceous, and is curiously laminated, the laminæ being in the direction of the main bedding and having a decided influence over the workable and merchantable character of the rock. The thinnest markings make the best cleavage and are divisions of the best stones. Mr. Gossip says also that "while looking at this excellent material for pavement, and considering the little difficulty there is in quarrying it, and its apparent abundance I could not but feel mortified at the spectacle exhibited in Halifax of a large importation of Caithness flag when an imperishable article, as good at least, if not much superior, may be found at our very doors."

Under the name of ironstone a quartzite, perhaps the same as the last, is largely used in building walls in Halifax.

Large quantities of coarse slate for pavements and foundations have for some time been quarried close to the railway station at Beaver Bank, 16 miles from Halifax; and on the shore about four miles east from Lunenburg an arenaceous rock probably of the same nature is got for the same purpose. Mr. Poole mentions the occurrence of slates for flags, under-pinning, and foundations at Clare and other places in the western counties, where large slabs may be obtained and good materials for various building purposes may be found wherever gneiss, mica slate and the more compact forms of clay slate appear in the southern metamorphic district of the province, and the slates of the newer metamorphic regions.

Roofing slates have been made to some considerable extent in Rawdon, and at the Gore, Douglas, Hants county; at the latter place the beds are vertical and slates of any dimensions, Mr. Lang informs me, can be obtained. Dr. Lawson has since reported favourably on the deposit as containing a variety of slates useful for various purposes; "the supply is not likely to be exhausted for several generations even if extensively worked."

In Digby county, Mr. Poole saw bands of good roofing slate at Avour's Head. I have seen a good specimen from near Weymouth.

Rev. Dr. Robertson informs me that on the south mountain near the line which separates the county of Annapolis from that of Kings there is a large quantity of excellent slate.

Variegated Clay Slate, Kings County. As a material suited for internal decorative purposes must be mentioned the beautiful soft clay slate which is found at Beech Hill, near Kentville, and is said to be abundant. It may be readily cut with a knife and furnishes surfaces most agreeably variegated with concentric bands of different colours in long oval patterns. It was discovered I believe by the late Dr. Webster, of Kentville, who first shewed me specimens.

Pencil Stone Under this name Rev. Dr. Honeyman submitted to me for examination a clay slate found in quantity, I believe, in Antigonish county, where it is much valued for the making of pencils for writing on slates.

Oven-Stone. The new red sandstone of Cornwallis is used in the neighbourhood of Kentville for building ovens, it exists in abundance and may be obtained with the greatest ease, it is sometimes called river-stone, from being found in convenient blocks in the Cornwallis river. It is very easily cut to any desired shape with an axe and answers admirably for the purpose to which it is put. The blocks are sold at Kentville at 50 cents each, when containing about two cubic feet of stone.

Stones and Materials for Grinding and Polishing.—Grindstones. At several of the places before mentioned as affording freestones for building some of the sandstones are found suitable for grindstones, this is especially the case at the Joggins in Cumberland county. In 1836 Dr. Gesner described the making of grindstones at the Bank Quarry at the South Joggins from sandstones consisting of minute grains of quartz united by an argillaceous cement into a compact rock capable of being split into tabular masses. Grindstones made of the finer varieties of this rock were then an important article of commerce with the United States. The smaller were the most valuable, but stones were made of six feet in diameter and a foot in thickness. In the census of 1861 a steam grindstone factory was given as one of the factories of the county. The principal localities in Cumberland for grindstone making are Seaman's Cove and Ragged Reef. The following statement from the census of 1861 gives at a glance the number of stones then made, in what counties, and their value :—

GRINDSTONES MADE IN NOVA SCOTIA IN 1860.

Counties.	Number.	Value in Dollars.
Halifax	16	14
Colchester	606	608
Cumberland	42706	40166
Pictou	1293	990
Antigonish	49	49
Guysborough	703	1106
Inverness	128	188
Richmond	4	6
Victoria	45	85
Cape Breton	47	58
Hants	893	818
Kings	1	1
Annapolis	3	3
Lunenburg	2	8
Totals	46496	44100

in the four remaining counties none were made.

At present grindstones are shipped from Pugwash by Messrs. McDonald, to the extent of about 100 tons annually, the value being 12 dollars per ton ; as mentioned previously, they are included with the freestones in the official returns so that we cannot get at the quantity sent away even : however, as regards the Joggins the export is chiefly, if not entirely, grindstones, and the value of stone exported thence during six years ending Sept. 30th 1866, was 63,620 dollars.

I have no information as to the relative qualities of the stones made but no doubt in the great number of beds from which they are taken varieties exist corresponding more or less closely with certain stones noted for particular applications, such as the Yorkshire grit used for polishing marble and copper plates for engravers, and the Sheffield blue stone, a fine grained stone used in finishing fine goods.

Millstones. Dr. Gesner mentioned in 1836 that millstones were then made from granite at White Point, Canseau, which was preferred to any other in the vicinity for stones used in grinding all kinds of grain. Boulders of granite are employed in making millstones in various parts of the province and the syenite of the Cobequids has been highly spoken of as excellent for the same purpose.

Honestones. A collection of hones was shewn in the Provincial Exhibition of 1854 for which a certificate of merit was awarded to Mr. Adam Bower, of Shelburne. Mr. Poole reports that honestone was found on Pleasant River, Queens county, and again at Whetstone Lake, in Shelburne county, where he met with loose pieces all along the south and west sides of the lake and also saw a piece apparently *in situ*. These stones are valuable when of particular quality and the deposits deserve investigation; a good idea may be obtained of the different kinds of hones in request from Hunt's Guide to the Museum of Practical Geology.

Burnishers. Agates and some kinds of jasper are used as burnishers; abundance of these minerals exists in the trap districts of the Bay of Fundy and Basin of Minas. At Blomidon a mass of agate of 40 lb. weight was obtained by Dr. Gesner. Several kinds of jasper are also found in the localities affording agates: very large quantities occur at Digby Neck, some of the varieties are said by Dr. Gesner to be very compact, these would be suitable for the purpose in question.

Cutting Material. The garnet sand found on the shores of Lake George, Shelburne county, used in the neighbourhood for dusting over the outsides of houses, consists of very small brilliant garnets of pale lilac colour. A similar mineral found near Pesaro, on the shores of the Adriatic, is employed for cutting hard stones, sawing marble, and like purposes.

Infusorial Earth. Deposits of the minute siliceous coverings of infusoria exist at Earltown and Cornwallis. The earth met with at the last named locality, discovered by the late Dr. Webster, of Kentville, is perfectly white, I believe it was found in some quantity. It was used very successfully by Messrs. Wesley and Sanford in polishing specimens of marble sent to the Exhibition in London in 1862.

There is practically no means of learning the amount of stone quarried in the province of late years but it must have been very large. The following table shews the value of the stone exported from 1860, before which year the returns were given under the head of "miscellaneous," to the last full financial year for which

L

returns were made to the local government. I have excluded all such entries as referred in whole or part to ores, quartz, and in fact to any "stone" but that apparently used for building and making grindstones.

VALUE OF STONE EXPORTED FROM NOVA SCOTIA IN THE YEARS ENDING 30th SEPTEMBER.

Ports.	1860.	1861.	1862.	1863.	1864.	1865.	1866.
Amherst				920			
Arichat				1,730		1,600	
Barrington				550	1,800		
Clementsport			250				
Halifax		29	575	841			
Hantsport						125	
LaHave			3,748				
Joggins	9,839	5,863	14,658	9,089	18,060	9,691	6,420
Pictou	6,084	4,395	1,740	2,205	5,380	3,155	2,135
Port Mulgrave					20	18	
Pubnico					354		
Pugwash		4		240	256	1,410	2,240
Tatamagouche	330	196	52	116	52	12	
Thorne's Cove		580					
Wallace	4,289	1,523	1,200	1,894	3,776	1,899	802
Totals	$ 20,542	12,590	22,223	17,585	29,698	17,923	11,667

The aggregate value for the seven years is 132,228 dollars.

CHAPTER XI.

MINERALS FOR JEWELLERY AND ORNAMENTAL PURPOSES NOT BEFORE MENTIONED.

So far as yet known the only minerals in the province available for jewellery and ornamental purposes other than those already spoken of are the different members of the quartz family, of which several interesting varieties are found, and the minerals topaz and garnet concerning which we have rather scanty information.

Topaz. Specimens of this stone were exhibited in London, in 1862, one piece was rough, the other cut and polished, the exhibitor was Mr. McDonald, the locality given was Cape Breton, and the cutting was reported as done in Pictou. The cut stone was about half an inch in length or rather more, its colour yellow. No information was given as to the precise locality of the mineral or the quantity in which it was found. The value of this stone and such others as are afterwards mentioned I give, unless otherwise stated, from the very interesting *Treatise on Gems* by Dr. L. Feuchtwanger, of New York, *(Third Edition,* 1867). Topaz is generally of less value now than formerly owing to the yearly supplies from Brazil which amount to about 40 pounds. The colours most esteemed are rose red and white. A topaz about the size of a bean is sold at Chapada, in the Termo Minas Novas, at one dollar, one of one carat, on the average, at eight dollars; yellow, three dollars; and a yellow burnt one, five dollars. In Brazil, very large fine and lustrous ones bring thirty dollars. The Saxon topazes are less valued; yet good yellow or crimson coloured stones, 9 lines long and 7 broad, bring 420 francs. The topaz is in general use by jewellers for setting in rings, pins, earrings, seals or necklaces; at present it is not a fashionable stone. Its fragments are ground and used in grinding the softer precious stones.

Garnet. In addition to the garnet sand before spoken of, garnets are found of small size in gneiss and mica slate in several parts of the province, especially in Shelburne county, as reported by Mr. Poole, who exhibited in 1862 a garnet crystal, of a deep red colour, three-quarters of an inch in diameter, which he had taken from a broken mass of gneiss rock lying on the surface; there were other crystals of about the same dimensions of which it was said they were too brittle to stand cutting and not of any intrinsic value. Those not exposed to the weather may be of superior quality. The value of garnets is determined by their degree of perfection, as well as colour, purity, and size. They are generally sold by the pound, containing from 60 to 400, valued at about eight or ten dollars; but a set of one thousand of the best selected garnets, well cut, is sold at about sixty dollars.

Quartz. Many beautiful minerals belonging to this family are found at numerous places in the trap districts especially along the shores of the Bay of Fundy and Basin of Minas. The action of surface water, frost, and the sea is continually changing the character of localities, specimens are brought to view, removed, and fresh supplies take their place. Dr. Gesner's description of the varieties of minerals and of some of the localities at which they were found in 1836 may be taken as a guide: the best marked varieties with their chief localities only can be specified here, the general list of localities, given as a subsequent chapter, can be consulted for further information.

Rock Crystal. Large geodes of crystallized quartz have been found but the mineral is not frequently met with in large crystals.

Amethyst. Purple quartz is found in bands and geodes at very many places, sometimes in considerable quantity. One of the best localities is thought to be Partridge Island, Cumberland county, where, however, the mineral has been removed so thoroughly that a fresh supply can only be got after a new fall of trap rock. It was from this spot that, as Dr. Gesner reports, DeMonts took some amethysts to Henry IV. of France, who was much pleased with them. At Cape Sharp, on the same shore, nearly opposite Blomidon, a geode was found described as capable of containing, before it was broken, two gallons. Its inner surface was studded over with large and regular crystals of a deep violet colour, over which was a light incrustation of siliceous sinter. Amethyst is found at many places on the opposite shore from Blomidon to Digby. Blomidon affords large masses often of deep colour, frequently in the form of geodes. Dr. Gesner says "it occurs in cavities in the amorphous trap; a single block presented a surface of a foot square perfectly covered with splendid crystals some of which measured an inch in diameter. Some of the amethyst found on this shore is seldom surpassed in beauty, a crystal from Blomidon is in the crown of the French King, and other pieces have been much admired in England and the United States." I have seen a band of amethyst of some feet in length and perhaps two inches thick, a mile or two east of Hall's Harbour, and a considerable quantity of fine mineral was got three or four years ago at Harbourville, also in Kings county. Some years ago Dr. Webster of Kentville had more than a bushel of good specimens found in digging a well in Cornwallis, in which township amethyst occurs occasionally with magnetic iron. Near Sandy Cove, Digby county, a geode of more than forty pounds weight was found by Jackson and Alger.

The amethyst is valued by the jeweller in proportion to the depth, richness, and uniformity of its colour; it forms when perfect a stone of exquisite beauty, its colour being perhaps more generally attractive than that of any other gem. The best amethysts now in commerce come from Ceylon, Siberia, and Brazil; the first are commonly called oriental amethysts which however must be carefully distinguished from a much more valuable gem, the true oriental amethyst, which is the violet sapphire. Good, well cut amethysts, such as now spoken of, of one carat are worth from three to five dollars, and so on in proportion to their size: an amethyst $1\frac{1}{4}$ in. long and nearly an inch broad, exquisitely fine, **was valued at 500 dollars.**

Smoky Quartz. This mineral, called Cairn Gorm in Scotland, where it is found in mountains of that name, is met with in several localities, of which the most noted are near Paradise River, and the neighbourhood of Bridgetown and Laurencetown, Annapolis Co. Immense crystals have been found here, some almost as transparent as glass, of a rich yellow colour, while others have the characteristic dark smoky appearance. I have seen a crystal, a six-sided prism, about thirteen inches in height and six in diameter, in the possession of Rev Dr. Robertson, rector of Wilmot: when at Paradise, some years ago, fine specimens were offered me at several dollars apiece. In 1836 Dr. Gesner said the crystals were easily obtained from the decomposing granite, but the great demand for them had rendered them scarce and they could not then be bought under their full value, while a few years before they had been piled up among the common stones of the field whence many had been taken to the United States. Crystals of 100 lb. weight are reported to have been found. Nichol mentions (*Mineralogy*, p. 111,) that when the cairn gorm was much esteemed a lapidary in Edinburgh cut £400 worth of jewellery from a single crystal. In the rapid and incomprehensible mutations of fashion this stone may again become a favourite, and no doubt there is a large quantity to be found in the district mentioned.

Smoky quartz is also found at Mill Village, Lunenburg county; and at Margaret's Bay, Halifax county, where it occurs in crystals of two inches in length, of which I have had specimens composed partly of chlorite at their summits, the lower part being transparent. The cabinet of King's College contains a specimen of exceedingly dark colour, almost black, such as is called morion, in crystals about half an inch across, from Little River, about five miles from Halifax.

Chalcedony. This mineral is found in many parts of the trap district before mentioned. Near Trout Cove, in Digby county, it is milk-white and of fine quality well **adapted for** seals and rings. Gesner found near the head of St. Mary's Bay in the same county, the peculiar kind of chalcedony called, from its appearance when polished, "cat's eye;" Partridge Island is also given as a locality by Marsh, who says the mineral is rare. This stone is much valued; of **the nearly opaque varieties,** the red and the almost white are most esteemed, and such are usually sold at from 10 to 20 dollars; a stone of an inch square, perfect in properties, is worth from 80 to 100 dollars. Carnelian is the name generally confined to the red varieties of the preceding, though it is sometimes given to those of white colour. The red coloured mineral is found at Blomidon, Kings county, at Trout Cove, Digby county, and the north shore of Granville, Annapolis county. Polished specimens from Blomidon are in the collections of the Halifax Mechanics' Institute.

Agate. This is the name given to mixtures of chalcedony and carnelian with other quartz minerals, such as hornstone, jasper, amethyst, quartz, heliotrope, cacholong and flint, and according to the predominating substances it is sometimes called chalcedony, jasper or carnelian agate. Many beautiful varieties are found in the trap regions of the province some of which are particularly described in Dr. Gesner's work so often mentioned. The agates on the shore extending from Sandy Cove to the head of St. Mary's Bay, Digby county, exhibit several varieties: among them is fortification agate, some are composed of alternate layers of transparent and white chalcedony, jasper, and quartz, curiously waved, and often lined crosswise with rays, sometimes jasper amethyst and chalcedony are united in such a way as to form brecica and dotted agate. These mixtures would be very beautiful in the polished state. Near Trout Cove are agates having a base of semi-transparent chalcedony, studded with irregular fragments of jasper and hornstone; sometimes the jasper is curiously striped with zigzag lines of red carnelian, forming a kind of agate not elsewhere observed. The agates are in veins of basaltic trap from half an inch to two inches in width. Near Blomidon large blocks of agate have been found, one described by Gesner, weighing upwards of 40 lb., consisted of semi-transparent chalcedony with curved fortifications of white chalcedony interlaced with lines of red carnelian. On one side a perfect onyx was formed. Another mass, of 80 lb. weight, was found a few miles east of Cape Split, which exhibited distinct and parallel zones of different colours: these zones consisted of circles of white cacholong with small rings of chalcedony and pale red carnelian; when polished these were extremely beautiful, and resembled the eyes of certain animals. Such agates are called eyestone. Sometimes there were white lines of cacholong and grey circles of chalcedony. Among other curious figures presented was one imitative of the gay figures of Indian quill-work. Beautiful moss agates are found at Two Islands, Cumberland county, and near Cape Split, and at Scot's Bay, Kings county. The last named place was considered by the late Dr. Webster, who was familiar with most of the localities, to afford the finest specimens: Partridge Island also and other places are localities of this variety.

Agate is used not only for jewellery and simple ornamental purposes, but for numerous useful objects, such as slabs, mortars for the analytical chemist, pestles, burnishers, handles of knives and forks, etc. Immense quantities of the mineral are worked up in Germany, for example, at Oberstein, in Rhenish Bavaria, there are five large manufacturing establishments where more than 100,000 dollars worth of work is turned out annually for export. The objects are very low priced, the best workmen receiving only a dollar and a half of weekly wages. Agate though much reduced in value as compared with former days still commands a tolerably good price: it is particularly onyx which is still at high prices. This variety consists of numerous layers, often as many as ten, of distinct and different colours and a certain thickness, running parallel with the larger surface, whereas the common ribbon agates display their various layers on the surface without being parallel. The predominating colours are usually greyish white, brown, and black. Sardonyx has one layer or more of carnelian, and this is the most esteemed. The finest cameos and intaglios of the ancients are of these materials, which have also been wrought into cups, urns, and other objects at enormous cost, in modern times.

Jasper. In the same localities as the preceding are found many varieties of jasper. It exists in very large masses on Digby Neck, where it presents red and purple striped, and red and yellow striped, varieties. Long Island has veins of red jasper. Near the head of St. Mary's Bay are large irregular blocks of red and yellowish red jasper, some of which is very compact. Some pieces are curiously striped with different colours, in others rounded pebbles of chalcedony are united by a siliceous cement. Large pieces of this breccia would afford an agreeable imitation of Mosaic pavement when polished. A large mass of Digby jasper was exhibited in Halifax, in 1862, which was much admired. Blomidon affords masses of jasper of several varieties, as does also Partridge Island, among which may be named a green from the former and a pale yellow with minute mossy markings of a black colour from the latter place.

Jasper is principally used in making seals, snuff boxes, vases, table-tops, and for architectural purposes. **It is worked at enormous cost in the Imperial workshops of Russia. It is hard to work and has no great value in commerce unless it is of exquisite quality.**

Heliotrope, or blood-stone was found in small nodules or fragments **of rock on the beach of Chute's cove,** Annapolis county, by Dr. Gesner, who says they may be picked out of **the sand where they accumulate** from the wearing away of the rock which had contained them. **It is leek-green** in colour with yellow and red spots. It is much admired; its value depends on the colour and quantity of the red spots: from one to twenty dollars is the usual price for **good and large specimens.**

Opal is mentioned by Gesner among the minerals of Partridge Island, two specimens were obtained in nodules resembling wax. It is well known that this is a gem of very considerable value when of good size and perfect character.

Semi-opal or common opal is found at a few localities: it is generally white or nearly so in colour, some of the mineral so called may be cacholong, which is white or bluish in colour and sometimes constituted of layers a quarter of an inch in thickness, of different degrees of hardness, alternating **with chalcedony.** (Feuchtwanger.) Generally, however, as found here, cacholong **is chalky white and quite soft, it frequently covers geodes of amethyst and agate externally**; it adheres to the tongue. The price of semi-opal is low, that of cacholong when of the character suitable for the lapidary is rather considerable. One specimen which I consider to be opal ,agate was found at Beech Hill, near Kentville, **Kings county; a portion** of it was given to Mr. Cornelius, Halifax, who had it cut and polished and the result was the production of handsome seal or ring stones composed of white and bluish white stripes about the sixteenth of an inch thick.

CHAPTER XII.

MINERALS NOT INCLUDED IN THE FOREGOING CLASSES AND CHIEFLY ADAPTED FOR THE CABINET.

EVER since Jackson and Alger described to the American Academy in 1833 the results of their expedition to Nova Scotia, made in 1827, on which occasion it is said they carried off a schooner load of specimens, the province has been celebrated among mineralogists, especially of North America, for the abundance and beauty of the minerals found in the trap districts before spoken of in describing the varieties of quartz. The collection referred to must, from what I have heard, have been very fine, it was only last autumn indeed that I was told "Alger's Nova Scotian specimens were magnificent; on his death his whole cabinet sold for some trifling ten thousand dollars or so, but it was really worth much more; his friends, however, were rich and perhaps thought the minerals were lumber." Numerous collectors from the United States as well as residents in the Province have since visited the localities and year after year, possibly without interruption, large quantities of fine specimens have been gathered. It is the object of the present chapter to give a brief account, a full one would be impossible in narrow limits and must be sought in treatises on mineralogy, of the most characteristic trap minerals and of others, suited for the cabinet of the mineralogist, which have not yet been mentioned in this Report. The fullest popular description of the trap minerals is given along with those of the quartz family by Dr. Gesner; since his book was written, however, some progress has been made by myself and Prof. Marsh, of Yale College, towards the accurate discrimination of the zeolites and other minerals and some new species have been distingushed as shewn in the papers on these minerals referred to in the Introduction. On this point

Prof. Marsh says, after stating that there is probably no part of the world, except the trap district of India, which is richer in zeolites than the shores of the Bay of Fundy, and that much confusion exists in regard to what species exist at the different localities, the entire group is well worthy of study and he has been for several years collecting materials for a full examination of the different species and he hopes at some future time to embody the results of his investigation in a monograph on the subject. (Mineral Localities, N. S., etc., Silliman's Journal, Jany. 1863.) A general list of the localities of the minerals of the province founded on that given in the paper just quoted forms a subsequent chapter of this Report, and may be referred to in connection with the following descriptions.

Analcime. Occurs in trapezohedrons, also massive granular. Colour white, occasionally grayish, greenish, yellowish or reddish-white, sometimes red; streak white, transparent or opaque, brittle. Fine crystals of this found in trap, sometimes an inch in diameter; often associated with natrolite. At Two Islands I have found large crystals partly consisting of chlorite lying loose in a cavity, above high water mark, like specimens on the shelf of a cabinet. I have one large crystal of which about half consists of chlorite.

Chabazite. Usual forms are rhombohedrons, which may be mistaken for cubes. Colour white and flesh red Streak uncoloured. Transparent-translucent, brittle. Very frequently found; often in beautiful crystals, sometimes much modified. The deep red variety is called acadiolite, probably from being found first in this province; (it is met with in Pennsylvania, T. D. Rand).

Ledererite or Gmelinite. Prof. Marsh has just shewn (Silliman's Journal, Nov. 1867.) that these minerals are identical: the specimens examined were found after many an unsuccessful search, at Cape Blomidon, nearly opposite Cape Sharp, in a locality supposed to be exhausted: they were in short hexagonal prisms with pyramidal terminations of from $\frac{1}{10}$ to $\frac{1}{3}$ of an inch in diameter. Some of them, especially the smallest, were colourless and nearly transparent; others were yellowish white or faint salmon red, and translucent. I think a specimen answering to this description was sent to Paris last year among the Nova Scotia minerals.

Mesolite. A mineral in columnar and radiated masses often mistaken for thomsonite which probably has not been found in the province in its common form. I first recognized mesolite here in 1858. It is found in large masses in the North mountains of Annapolis in glassy and in opaque crystals. It is a variety of "needlestone."

Natrolite can only be distinguished from the foregoing by chemical analysis when in closely compacted crystals. It however, often forms bunches of nearly separate transparent crystals alone or with analcime, its crystals are sometimes hair-like and beautifully silky.

Faröelite. This mineral I detected here in 1859 in its common association with mesolite. It occurs in hemispherical masses underlying the mesolite, that is, next the matrix, and may at once be distinguished by its pearly appearance. At Bishop's Brook, near Margaretville, I obtained, in 1863, masses of this mineral shewing the termination of the crystals. (Dana now considers this to be a variety of thomsonite. *Mineralogy, Fifth Edition.*)

Heulandite. Oblique rhombic prisms with modifications. Colour various shades of white passing into red, grey, and brown. Streak white. Transparent — subtranslucent. Brittle. This beautiful mineral occurs in great abundance, sometimes in crystals an inch and a half in length. It frequently lines geodes, one of these I saw in the possession of the late Dr. Webster, of Kentville, about nine or ten inches in height and nearly half as wide, its walls were about an inch thick, the interior was covered with large, brilliant colourless crystals of heulandite. The specimen was much valued by Dr. Webster and much coveted by his visitors; found by a companion of the Doctor's on a mineralogical expedition it passed to him on the death of the latter according to arrangement.

Stilbite. Common in sheaf-like aggregations, globular, divergent, and radiated forms, also thin lamellar and columnar. Its colour is white, occasionally honey yellow, brown or red. Streak uncoloured. Subtransparent—translucent. Brittle. This is perhaps the best known of the zeolitic minerals. It is found in very many localities, and not seldom in masses of very large size beautifully crystallized on the surface: these masses sometimes weigh hundreds of pounds.

Apophyllite. In foursided prisms, with pyramidal terminations: crystals sometimes nearly cylindrical or barrel-shaped. Colour white or grayish, occasionally greenish, sometimes of a beautiful green, also yellow or red. Streak white. Transparent—opaque. Brittle. Distinguished by its white pearly aspect resembling the eye of a fish after boiling. Not so common as the two preceding and not very frequently met with in fine crystals, but very beautiful at certain localities, especially Isle Haute. I have seen very large crystals, an inch and a half across, at Port George, Annapolis county. An American collector offered the late Professor Chipman, of Acadia College, who unfortunately was drowned in his zeal for collecting trap minerals, two sovereigns for a specimen of apophyllite about the size of the palm of the hand, but in vain.

Gyrolite. Occurs in spherical concretions having a lamellar radiated structure, white and pearly Originally found some years ago in Skye, it was detected here in 1861 by myself. It is a rare mineral, met with on apophyllite and some other minerals.

Centrallassite. This is a mineral very like the preceding and close to it in chemical composition, which I described as new in 1859. It appears to me that the mineral described by Gesner as Prehnite (*Remarks*, p. 202) was either one or the other of the two preceding.

Cyanolite. A bluish mineral somewhat resembling chalcedony found with centrallassite, and with one of a yellowish colour resembling wax, described as also new by myself under the name of *cerinite.* (Paper on *Three New Minerals* referred to in Introduction.)

Mordenite. A mineral of fibrous structure in hard nodules in trap, described as new by myself in 1864. On shewing specimens of this to C. F. Hartt, Esq., Professor at the Cornell University, Ithaca, N. Y. State, who has often collected minerals here, he at once recognised its individuality. Prof. C. U. Shepard, of Amherst, Mass., to whom I sent specimens at his request, sent me two minerals, one from Skye, Scotland, the other from India, which he thought might be the same; the former had been sent to him as a new mineral, both were in too small quantity to examine chemically.

Laumonite. Columnar, radiating, and divergent. Transparent—translucent. Becomes opaque and usually pulverulent on exposure: hence known as efflorescent zeolite. Colour white, passing to yellow or grey, sometimes red : sometimes green from salts of copper. Abundant in several localities especially in Annapolis county, in veins some inches in thickness running down the whole face of the trap rocks. Specimens may be preserved for some time by dipping them in gum water.

Chlorophœite. Foliated or granular massive, imbedded. Colour dark green, often changing to black.

Green Earth. Earthy or of minutely crystalline appearance, dark olive green to celandine-green, and quite soft, with an **unctuous** feel. Occurs in cavities in amygdaloidal trap.

Poonah Earth. A bright green earthy mineral in cavities of trap and on certain zeolites. The three preceding are sometimes taken as indications of copper which metal is not present in **either** of them.

Obsidian or Volcanic Glass. This is distinguished from quartz of the same colour by its fusibility before the blow-pipe.

Siliceous Sinter. This is the name given to irregularly cellular or porous quartz formed by deposition from water. Here it is often misapplied to opaque quartz, of a milk white or pink colour, in distinct and beautiful crystals.

The preceding are all trap minerals; there are a few others met with in the same districts, as noted in the list of localities, which do not call for special remark. Among the other minerals it will not be necessary specially to notice any but the few following otherwise than in the general list.

Wichtisite. A very rare mineral brought to me by a farmer from Cornwallis, in which township I understood it to have occurred. It is known only at one other locality, namely at Wichtis in Finland. I was told it had received the name of "the little pebbles," the specimens brought me were something like obsidian, of grey and deep blue colours. (See *Contributions to Mineralogy of N. S.*, II., referred to in the Introduction.)

Mr. Barnes informs me that perfect crystals of *flesh red felspar*, sometimes 8 inches in length, are found in syenite, at Cheticamp, Inverness county, C. B., also, that he has found *Apple Green Calcite* of great beauty, and *octahedral fluor spar* of blue colour in crystals more than an inch across in the same county; *anatase* (titanic acid,) in small but fine crystals, in quartz, at Sherbrooke, and fine *chiastolite* in slate at Chaplin's saw mill in the north-east of Halifax county.

CHAPTER XIII.

MINERAL WATERS.

In addition to the brines before described many mineral waters in different parts of the province have attracted attention. Although several have long had the reputation of possessing considerable medicinal virtue, and some have undoubtedly been found of great value, few of them have been examined chemically. I have made detailed analyses of three or four of the best known (see papers on Waters referred to in the Introduction) and partial examinations of others: besides the results so obtained there exists very little information on the subject. Such as is known to me is given in the following account which contains also such general descriptions of the localities and nature of the waters as I have been able to furnish from observations made by myself and from sources specially mentioned. It will appear that the waters present varied and interesting characters.

Saline Water of Bras D'Or, Cape Breton. This is the most remarkable water in the province, as regards chemical composition, so far examined. It had at the time I analyzed it (1859) an extraordinary and well grounded reputation as being very efficacious in various maladies, authentic cases being known of much benefit resulting from its use in rheumatism and severe headaches. A gentleman of high standing and scientific reputation informed me that he had obtained a good appetite and increased strength by taking about five gallons of it and by further use a moderation of the violence of asthmatic attacks to which he was subject, in fact that its employment had been a real blessing to him. A water possessing such qualities would of course be much resorted to and it was considered worth while to erect a house for the accommodation of visitors soon after its merits became somewhat known.

There are three springs mentioned as affording the water examined; they are situated near Kelly's, on the high road from Sydney to St. Peter's, in a brook which empties into the Salmon River and is distant about two or three miles from the source of the river and six or seven from the southern shore of Bras d'Or lake. The waters rise in syenitic rocks and the flow is not more than a gallon per minute. The amount at my disposal did not enable me to make the most perfect examination possible, so that the following results do not express with rigid accuracy the composition of the water; they are calculated for the imperial gallon of 70,000 grains. The water was clear and of neutral reaction; it afforded:—

Grains in a gallon.

Iron and phosphoric acid	traces.
Carbonates of lime and magnesia	0.60
Sulphate of lime	0.94
Chloride of sodium	343.11
Chloride of potassium	4.55
Chloride of calcium	308.90
Chloride of magnesium	4.47
	662.57

Specific gravity at 54° Fah..........1,007.397

No iodine was detected in the residue left by 1,500 grains of the water. The composition brought out is very remarkable and at the time my analysis was made the only similar waters found were those from certain springs in Canada (as it was then) described by Dr. Hunt in Geology of Canada, 1863, p. 531. The great feature in the case is the large proportion of chloride of calcium and the small quantity of sulphate present. Since the date of my analysis English waters have been examined approaching the foregoing in composition (Chemical News, X. 181 and XIV. 244) but the closest resemblance is still with those of the upper provinces forming the first of the six classes into which Hunt has thrown the mineral waters occurring there; they are characterized as "containing chloride of sodium with large portions of chloride of calcium, sometimes with sulphates. The carbonates of lime and magnesia are either present in very minute quantities, or are altogether wanting. These waters are generally very bitter to the taste and always contain portions of bromides and iodides." It is further stated that

M

such springs are altogether unlike any studied and may be supposed to represent the composition of the ancient ocean in which the early strata (of lower silurian age) from which they originate were deposited.

At Cheticamp, Inverness County, C. B., a spring issues, Mr. Barnes informs me, from lower carboniferous limestone with an oily matter on its surface having a strong odour of petroleum : the limestone has the same peculiar smell. At the copper mine, before mentioned, at the same place, a spring affords water highly impregnated with sulphate of copper; it flows from the mountain. At Grand Anse, at the mouth of the M'Kenzie River, in the same district two springs issue from the metamorphic lower carboniferous rocks resting on the flanks of a mountain of granite and syenite. The first is highly sulphurous and contains sulphate of magnesia, it strongly resembles the sulphur springs of Harrowgate and the water has very decided aperient qualities. The little pool in which it rises is coated with a white earthy deposit : gas is evolved, particularly when the neighbouring ground is trodden on. The second water is mentioned as having a strong taste of magnesia, not having any sulphurous odour, and as being much used as a gentle laxative. With regard to the water issuing in direct contact with the gypsum in the plaster districts of Cape Breton, Mr. Barnes says they always appear tolerably pure with the exception of being rather hard from presence of sulphate of lime.

Thermal Spring near Chester, Lunenburg County, N. S. Mr. Amos F. Morgan has furnished me with the following account of a spring of clear water issuing in the centre of a small hillock in the woods near Chester. When the spring was found, in March, the temperature of the air was below freezing, while the water was about as warm as new milk and the pool had no appearance of having been frozen over. The basin filled with the water was estimated to be about eight feet square, the mud at the bottom was full of small holes from which gas was continually rising. The water was peculiarly soft so that it felt more like oil than water in the mouth, it was thought also to be slightly bitter. From the description given the water is probably of the alkaline class.

Spa Spring Water, Windsor, Hants County. This water rises in a field, on the grounds of C. B. Bowman, Esq., close to the Forks

Road: gypsum is one of the prevailing rocks of the district, the geological age of which is lower carboniferous. The water has long been considered chalybeate and is still sometimes taken medicinally. It is well known to be a very favourite drink with horses and cattle. The chalybeate character of the water was inferred from a certain red deposit found in the pipes through which it was made to run, and from its strongly inky taste. The iron to which these effects are due, however, does not exist in the water as it issues from its outlet, as is shewn by the following analysis made on water carefully collected by myself in a small reservoir filled immediately from the spring beneath. The water was perfectly colourless and clear, it had little taste and that not inky; its temperature was $49°$, that of the air being $31°$, Fah., (it flows all the winter). It gave the following ingredients to the imperial gallon, in December, 1858:

	Grains in a gallon.
Carbonate of lime	17.50
Carbonate of iron	0.40
Carbonate of magnesia	0.31
Sulphate of lime	106.21
Sulphate of soda	0.68
Sulphate of potassa	0.38
Sulphate of magnesia	11.02
Chloride of sodium	0.90
Phosphoric acid and organic matter	traces
Silica	0.60
Total	138.00
Free carbonic acid (1.35 cubic inch at $33°$)	0.64
Specific gravity at $49°$ Fah	1,001.858

This water would be placed in the sixth class of Hunt from its richness in sulphates (Geol. Canada, 1863, p. 532). The sulphate of lime, which is the characteristic ingredient, is present in larger amount in one only of the fifteen waters of Cheltenham, in England, and is by no means a common constituent of waters in such large proportion. The water is known to possess aperient qualities when taken freely. The inky taste and red deposit above spoken of are due to its action on the soil through which it is allowed to flow and to its admixture with surface water, and the impregnation thus obtained, only observed when precautions are not taken to keep the spring water pure, is of course subject to variation.

Water of Wilmot Springs, Annapolis Co. About 40 years ago this water was in high repute, as an abridgement from Dr. Gesner, writing in 1836, will shew. " In the township of Wilmot about 3 miles from Gibbons's, on the high road to Annapolis, there is a mineral spring possessing medicinal properties of considerable importance. When the discovery was first announced numerous persons, without reference to the nature of their diseases, and at every stage of their complaints, hastened to the waters and hoped, and vainly hoped, to obtain relief. The little village near the pool was all bustle and confusion, while many for want of accommodation were obliged to depart and few of the requisite comforts and conveniences could be procured for those who remained. Many and great were the cures reported but experience shewed that the powers of the waters were not sufficient to remove all the ailments of its visitors; hence the springs were soon abandoned, but were they surrounded by the pleasing scenery and cheerful society of the European watering places, the waters would be found at least equal to many at the celebrated resorts: and even now the time is not far distant when the public mind will react and many find relief at the deserted pool. The waters of the spring have been analysed by Dr. Webster, of Boston, and were found to contain iodine, lime, sulphuric acid and magnesia. They will doubtless be beneficial in all scrofulous and glandular diseases, and probably in the first stages of tubercular consumption. They are gently aperient and cannot fail to be serviceable in dyspepsia and other diseases of the digestive organs. This might be supposed from their composition and has been confirmed by experience."

I visited the springs in 1855 and copied from the register of visitors an analysis by Dr. Webster, no doubt the same as referred to by Dr. Gesner, which will be mentioned again. I paid a second visit in 1863, on which occasion I collected some water from two basins and analysed them with results which will be given immediately. When engaged in writing my account of the water I asked the Rev. Dr. Robertson, rector of Wilmot, for information on the subject and received in reply a letter containing the following: " No correct analysis of the water, I rather think, has as yet been made. It is said to contain a small proportion of iodine. In former times the springs were much resorted to but of late years very few visitors have been near them. The water, however, is remarkably efficacious in curing cutaneous complaints or eruptions. In my own opinion the Wilmot springs deserve to be better known and more frequented than they are at present. If the proprietors were men of substance and energy I have not a doubt that this locality would be one of the best known in Nova Scotia." On the ocasion of my recent visit I was driven to the spot by a very intelligent young man belonging to Margaretville, a village about 5 miles from the springs, who told me that people had been in the habit of coming from the United States and residing for months at Wilmot on account of the springs, and that the water had been frequently exported to order to the States or New Brunswick or both, also, that he had himself drunk the water for a twelvemonth and though he found it rather unpleasant at first he came to prefer it to any other; the effects were described as being decidedly purgative on the first use. When at the springs I had evidence that the water was still believed in for I saw a young man who had just finished applying to his leg, which had been badly cut with an axe, a thick coating of mud from the spring as a healing plaster. Several bathing houses were seen conveniently arranged, but apparently not much used.

I found the springs situated under lofty trees, a few feet off an excellent road, filling two basins. One of these was about six feet in diameter, affording a considerable overflow in a rapid stream about an inch in diameter running from a trough found very convenient for collecting the water. The other basin was at a distance of three or four yards and was not furnished with a trough, whence I concluded that it is not, generally at any rate, made use of, nevertheless, thinking it possible the two waters, though so close at their outlets, might be dissimilar, I collected at this basin also. The waters were beautifully clear, cold, and without smell or particular taste. It is proper to mention that the summer of 1863 was very wet and that heavy rains had fallen for a day or two before my visit which however had entirely ceased for seven or eight hours when I collected the waters. How far this rain fall would affect these copious springs would only be shown by the results of analysis made in a dry season for the sake of comparison. On proceeding to analyse the waters some weeks after collection, that from the larger basin was observed to have undergone alteration from the presence of organic matter, it smelt of sulphuretted hydrogen and some sulphate of lime was deposited: the results of the analyses of both were as follows:—

	Grains in a Gallon.	
	Larger Basin.	Smaller Basin.
Lime	54.69	51.74
Magnesia	2.74	2.94
Soda and potash	6.00	4.65
Sulphuric acid	78.03	79.07
Chlorine	1.09	.76
Silica	0.70	.55
Phosphoric acid	traces	traces
Organic matter	traces	traces
Carbonic acid	undet.	undet.
Iron	traces	(oxide) 0.09
	143.25	139.80

which shew that the waters are essentially the same.

As the water from the smaller basin had remained almost entirely unchanged the results it gave were properly calculated and arranged, and they stand thus; the carbonates were determined by boiling:

CONTENTS OF THE WATER IN 70,000 GRAINS.

	Grains.
Carbonate of lime	2.70
Carbonate of magnesia	0.37
Carbonate of iron	0.14
Sulphate of lime	121.98
Sulphate of soda	8.35
Sulphate of magnesia	5.35
Chloride of potassium	1.60
Silica	0.55
Phosphoric acid	traces
Organic matter	traces
	141.04

I failed to detect iodine in the residue of 7,000 grains from each basin. The resemblance of this water to that of Spa Spring at Windsor is apparent. The analysis before referred to as copied by myself from a book at the Springs where it was attributed to Dr. Webster, of Boston, was as follows, calculated for the imperial gallon:—

	Grains.
Carbonate of Soda	6.39
Carbonate of lime	1.92
Carbonate of magnesia	1.10
Carbonate of iron	.75
Iodine	3.06
Sulphate of soda	3.47
	16.69

These results are entirely different from mine, and there must either be some mistake about them or the water must have undergone one of those radical changes which have been observed from time to time in mineral waters in different countries. The quantity of *iodine* given is enormously large; thus in the water richest in iodine mentioned by Dr. Hunt, in Geology of Canada, the amount of *iodide of sodium* is under one grain to the gallon, and the iodine can be detected in the water itself without evaporation while I failed to find it in the residue of 7,000 grains of each of the Wilmot waters as mentioned above. With respect to the changes which waters undergo several are described by Dr. Hunt in the report on the Geology of Canada, 1866; for the most part they consist in the reduction of the amount of sulphates and the increase of that of carbonates, especially of alkalies, or just the reverse of what must have occurred if there has been change in the Wilmot water. The Harrowgate waters in England have also been found to undergo changes similar to those observed in the Canadian waters in question. Several of the Harrowgate waters, all of which were found by Dr. Hofmann in 1854 to contain sulphate of lime, were examined by Mr. Davis in 1866 and proved with one exception to be free from sulphate and to contain instead salts of baryta even in sulphuretted waters. (Geology of Canada, 1866, p. 279). From a recent analysis of the Montpelier saline chalybeate spring at Harrowgate, by Dr. Muspratt, it appears that the water was much stronger in 1867 than it was two years previously as regards its saline constituents, and that it had acquired chloride of barium, etc., while many years before it had contained as much as 20 grains of sulphate of soda to the gallon, which of course could not be present at all with salts of barium. (Chemical News, XV, 244). As remarked by Dr. Hunt, often repeated analyses are of course necessary to determine whether the changes in water are permanent or whether they are periodical and dependant on the changes of the seasons.

Other Mineral Waters. A water with a strong acid reaction is reported to exist near Gair Loch, Pictou county. Springs, locally famous, are found at Earltown, Shubenacadie, Hants county, and at a place about a mile and a half east of the county town of Shelburne. A sulphur spring is reported to exist at Cranberry

Cove, Cole Harbour, Halifax county. The water supplied to the city of Halifax, contained in the gallon, according to the analysis of Mr. Rickard, in September 1862, inorganic matter, 0.70 grains, consisting chiefly of chloride of sodium with traces of sulphuric acid, magnesia, and iron, and 1.65 grains of organic matter, a quantity of such impurity pointed out to be large and as arising no doubt from decomposing vegetable matter in the lake whence the supply is taken: the earthy matter was remarkably small in amount.

CHAPTER XIV.

CATALOGUE OF LOCALITIES OF MINERALS.

The following list contains the principal (when very numerous) and in some cases all the known localities of valuable and interesting minerals in the province. As regards the minerals of the trap districts the localities are chiefly those given by Prof. Marsh in his "Catalogue of Mineral Localities in New Brunswick, Nova Scotia, and Newfoundland." (Silliman's Journal, Jan. 1863.) This catalogue the author describes as "the first that has been published and, though necessarily imperfect in many respects, has been prepared with considerable care. The list of minerals occurring at many of the places mentioned (in N. S.), especially those in the trap district of the Bay of Fundy, are copied from the writer's notes taken at the localities during several excursions to the provinces, the first in 1854. Even these lists may, in some cases, be found imperfect since the destructive tides of that region are continually changing the outlines of the coast and thus exhausting the old localities, but at the same time bringing to light others equally rich in mineral treasures. The notices of localities which the author has not visited are derived from the best sources of information to which he had access. A few were taken from the publications of Jackson and Alger, and Dawson, which contain much that is valuable with regard to the mineralogy of the province. The author is also indebted to F. C. Hartt, Esq., of St. John, for important information in regard to localities." Reference is then made to doubts as to the existence of thomsonite and prehnite to which I have already adverted and finally the plan of the catalogue is thus stated. "The catalogue is arranged according to the plan used in Dana's Mineralogy. Only localities which afford cabinet specimens are in general included. The names of those minerals which can be obtained in good specimens are printed in italics. When the specimens are remarkably good an exclamation mark (!) is added, and two of these are given (!!) if the specimens are unique."

In making additions to the list of Prof. Marsh the same indications are employed in similar cases, and as regards economic "minerals" the names of localities, or sometimes of the substances themselves, are put in italics where there is known or believed, on good grounds, to be workable quantity. My own observations and the sources before mentioned in this Report have furnished most of the information now for the first time put together. Frequent communication with the late enthusiastic collector of specimens Dr. Webster, of Kentville, with whom I had the privilege of visiting some localities in the Bay of Fundy and whose collections I have often examined, familiarized me with several interesting particulars, others have been obtained from Prof. F. C. Hartt, formerly of this province, where he collected minerals for a series of years. My experience in connection with the Exhibitions before spoken of has been of essential service. I am indebted to Mr. Barnes for several very interesting facts recently communicated.

NOVA SCOTIA PROPER.

CUMBERLAND Co. Amherst.—Pyrolusite (of fine quality.) Minudie to Pugwash and Wallace.—*Limestone, freestone, grindstone, gypsum;* French River, *vitreous copper ore* in sandstone: Napan River, *gypsum, limestone;* Macan R., Hebert R., Springhill, *coal.*

Joggins.—*Coal,* hematite, limonite, malachite, tetrahedrite at Seaman's Brook; galena, (in octahedral crystals), *grindstone, bituminous limestone.*

Cape D'Or.—*Analcime, apophyllite!!* (large crystals highly modified), chabazite, faröelite, laumonite, mesolite, natrolite, *native copper,* red copper (rare), malachite, obsidian, gold. Horse-shoe Cove, analcime, calcite, stilbite.

Spencer's Island.—*Native copper,* siliceous sinter, jasper, quartz (in perfect crystals).

Cape Sharp.—Stilbite, amethyst, magnetite, calcite.

Isle Haute.—South side, analcime, *apophyllite!!* albin? calcite, *heulandite!!* natrolite, mesolite, *stilbite.*

Parrsborough.—Augite, amianthus, calcite, gypsum, hematite, iron pyrites, magnetite, quartz, wad, earthy plumbago.

Partridge Island.—Analcime, *apophyllite!* (rare), *amethyst!* agate, apatite (rare), *calcite!!* (abundant in large and highly modified crystals, often straw yellow), chabazite (acadialite), chalcedony, cat's eye (rare), gypsum, hematite, *heulandite!* magnetite, *stilbite!!* (very abundant), jasper, cacholong, opal, semi-opal, gold in quartz.

Clarke's Head.—Analcime, anhydrite, chlorite, calcite, hematite, (specular ore), prehnite? tremolite, pseudomorphous quartz after stilbite.

Swan Creek.—West side, near the Point, calcite, gypsum, *heulandite*, iron pyrites; east side, at Wasson's Bluff and in vicinity, *analcime!!* (occasionally enclosing native copper and malachite), *natrolite!! apophyllite!* (rare), calcite, chabazite!! (white, wine yellow, red, [acadialite] in large and very perfect crystals), gypsum, *heulandite*, malachite, native copper, red copper (rare), siliceous sinter.

Two Islands.—Moss agate, *analcime* (sometimes in crystals partly chlorite, lying loose in cavity of trap), calcite, chabazite, *heulandite*, selenite.

M'Kay's Head.—Analcime, calcite, heulandite, *siliceous sinter!*

Stronix Brook.—Laumonite.

KINGS COUNTY. Cape Blomidon.—On the coast between the Cape and Cape Split the following occur in many places, some of the best localities are nearly opposite Cape Sharp, *analcime!!* agate, *amethyst! apophyllite!* calcite, chalcedony, chabazite,- gmelinite (ledererite) (rare), faröelite, hematite, magnetite, *heulandite!* laumonite, fibrous gypsum, malachite, *mesolite,* native copper (rare), *natrolite!! stilbite!* psilomelane, thomsonite? *quartz.*

Scot's Bay.—*Agate* (especially moss), amethyst, *chalcedony,* mesolite, natrolite.

Woodworth's Cove.—A few miles west of Scot's Bay, *agate,* chalcedony, jasper.

Hall's Harbour.—Amethyst, centrallassite, *red heulandite, stilbite.*

Harbourville.—*Stilbite, heulandite,* amethyst.

Black Rock.—Centrallassite, cerinite, cyanolite, flesh red heulandite, laumonite; a few miles east, *calcite!* prehnite? *stilbite!* agate.

Long Point.—Five miles west of Black Rock, *heulandite, laumonite!! stilbite!!*

Morden or French Cross.—Mordenite, *stilbite!*

North Mountains.—Amethyst, bloodstone (rare), *ferruginous quartz*, mesolite (in soil).

Cornwallis.—At the bridge, manganese (in conglomerate); mouth of river, Long Is., Boot Is., *sandstone, ovenstone;* Canning? *wichtisite!* Aylesford, *peat-bog.*

Greenwich.—Manganese (pyrolusite and psilomelane).

Little Chester, and several places on South Mountain.—Gold.

Beech Hill.—*Bog iron and manganese* (exported for pigment), opal-agate (loose), slates for roofing and others beautifully variegated and soft.

New Canaan.—Slates.

ANNAPOLIS COUNTY. Chute's Cove.—*Apophyllite,* natrolite.

Gates's Mountain.—Analcime, magnetite, *mesolite! natrolite!* stilbite, thomsonite? Hadley's Mountain.—Chlorophœite, heulandite.

Margaretville or Peter's Point.—Laumonite (abundant, some coloured green by copper), stilbite; west side of Stonock's Brook, *apophyllite!* calcite, heulandite, native copper, stilbite; near the pier, *analcime,* mordenite, sometimes with gyrolite; near Bishop's Brook, about three miles east of pier, native copper (in crystals in vein of zeolite).

Marshall's Cove.—*Analcime* (enclosing native copper), chabazite, *heulandite.*

Moose River.—*Beds of iron ore, moulding sand,* clay.

Nictau River.—At the Falls, *bed of hematite;* west of Falls, on hill, bed of *magnetic iron ore;* in the district, magnetic iron pyrites.

Paradise River.—Black tourmaline, *smoky quartz!!* (perfect crystals, more than 100 lb. in weight, have been found in the soil.

Port George.—Farœlite, laumonite, mesolite, stilbite, *apophyllite* (in large crystals); east coast, gyrolite in apophyllite.

St. Croix Cove.—Chabazite, heulandite; between this and Chute's Cove, heliotrope.

Wilmot.—Water of the Springs medicinal; at the Springs, copperas?

Near Annapolis.—Clay-beds (brickyards).

DIGBY COUNTY. Clare.—Avour's Head, gold in quartz, talcose slate, garnets in chlorite slate, arsenical pyrites; Meteghan, Indian pipe-stone.

Briar Island.—Native copper in trap, jasper, magnetite.

Digby Neck.—Sandy Cove and vicinity, *agate, amethyst, calcite, chabazite, hematite,* (in perfect crystals), micaceous iron, *laumonite* (abundant), magnetite, *stilbite,* quartz crystals, cat's eye, jasper (abundant, yellow and red, striped); gold in quartz? Gulliver's Hole, magnetite, stilbite. Mink Cove, magnetite, amethyst, *chabazite!* (crystals an inch in diameter), quartz crystals. Trout Cove, six m. east of Sandy Cove, *agate, chalcedony.* Sea-wall, **specular iron ore** at Johnson's.

Nichol's Mountain.—South side, *amethyst, magnetite!* (in large and perfect crystals). Cowan's, jasper, agate, iron ore. Timpany's, *hematite,* amethyst, ferruginous quartz. Peters's, magnetic iron ore.

Williams's Brook.—Near the source, chabazite (green), heulandite, stilbite, quartz crystals.

Marshalltown and Bear River.—Gold.

YARMOUTH COUNTY. Tusket.—Crosby, 1½ mile from town of Tusket, lead ore, arsenical pyrites, in quartz.

Yarmouth.—Jebogue Point, copper pyrites ; Cat Rock, Fourchu Point, calcite, asbestus ; Foot's Cove, garnets in chlorite slate ; Crannbery Head, smoky quartz ; Cream Pot, gold in quartz.

SHELBURNE COUNTY. Shelburne.—Near the town, pebbles of rose quartz ; near Birchtown Bay, peat bog ; Jordan and Sable Rivers, staurotide [abundant in gneiss and mica slate], schiller spar ; Stokes's Head, graphic granite, garnets ; **Port Herbert,** red ochre, garnets in gneiss ; Kail's **Point, green quartz [six feet]** ; Shelburne Road and Wharf, garnets in gneiss ; Falls, granite, with *mica* [in large plates] ; Fifteen miles up the river, east side, bog iron ; Indian Fields, talcose slate ; Whetstone Lake, honestones ; near Clyde River, *peat bog.*

Barrington.—Clyde River, tourmaline ; Fresh Pond, vein of felspar ; Port LaTour, andalusite, bog iron ; Upper Pubnico, andalusite ; Argyle, slates for underpinning.

QUEENS COUNTY. Liverpool.—Five Rivers, near Big Fall, gold in quartz, smoky quartz, ferruginous quartz, bog iron ; Little Port Joli, smoky quartz.

North Queens.—Westfield, gold in quartz, rose quartz, copper pyrites in chlorite slate ; Pleasant River, honestone ; Harmony, slate, [hard and strong] ; Hibernia, limestone [loose] ; Brookfield, limestone [loose] ; Hillsborough, white iron pyrites.

LUNENBURG COUNTY. Bridgewater.—LaHave, *iron pyrites* [large crystals]; Hebb's Bridge, bog iron ore; Hebb's Road, steatite; Mill Village, smoky quartz; Three Mile Lake, azurite in slate; Lapland, Seaman's Farm, bog iron ore; Petite Riviere, gold; Indian Brook, gold; Branch Lake, gold.

Chester.—Gold River, *gold* in quartz and sand, *argentiferous galena* in quartz, molybdenum; Chester Basin, *cement* and *paint stone, umber.* Near Chester, thermal spring.

Lunenburg.—The Ovens, Cross Island, Long Island, and other places not named, *gold* in quartz and washings, *pyrites, mispickel!* [in perfect crystals occasionally]; near Lunenburg town, hornblende; Waterman's Lake, vicinity of, manganese; Ritchie's Cove, ferruginous quartz.

HANTS COUNTY. *Renfrew, Uniacke,* Birch Bark Lake, Ellershouse, Stillwater, Ponhook? River Hebert, gold, etc., in quartz.

Uniacke.—Sulphur [rare, crystals]; Douglas, manganese; Rawdon, *slates.*

Cheverie.—*Anhydrite, gypsum, manganite,* pyrolusite.

Walton.—*Gypsum, manganite,* pyrolusite, hematite, near the bridge, brine spring; Anthony's Nose, barytes, [translucent crystals], red chalk.

Pembroke—*Pyrolusite,* manganite, *calcite,* barytes.

Teny Cape.—Pyrolusite! manganite! [abundant, especially former, both often in fine crystalline masses], *nail-head calcite,* dog-tooth spar, barytes.

Noel.—*Gypsum,* limestone. Maitland, *gypsum.*

Shubenacadie River.—*Gypsum.*

Kennetcook.—*Freestone, grindstone, limestone, gypsum,* coal [thin seam].

Newport.—*Gypsum,* anhydrite, natroborocalcite, silicoborocalcite, selenite, [last three in gypsum]. Parker's Mills, Meander R., pickeringite.

St. Croix.—*Gypsum,* red crystalline gypsum, *anhydrite,* pipeclay.

Falmouth.—*Freestone, gypsum, anhydrite;* west arm of Avon, barytes.

Windsor.—*Gypsum, anhydrite,* fibrous gypsum, selenite, birdeye gypsum [these varieties at other localities also], calcite, natroborocalcite, cryptomorphite, glauber salt [perfect crystals], common salt [preceding associated]; *limestone, marls, moulding sand,* Spa Mineral water. Fall Brook, sandstone.

Brookville.—Three miles south of Windsor, *gypsum* holding natroborocalcite, silicoborocalcite, arragonite; *anhydrite* holding silicoborocalcite, arragonite or calcite; selenite.

Seven Mile Plain.—Iron pyrites [large and often modified smaller crystals].

HALIFAX COUNTY. Halifax.—Vicinity of, native antimony? Harrietsfield, schorl; south-west of city, garnet, staurotide; Northwest Arm, *granite, flagstone*; Clewly Road, gold in quartz; Bedford, *iron pyrites!* Fenerty's, manganese; Five Mile River, smoky quartz [black]; Margaret's Bay, smoky quartz with chlorite. *Lawrencetown, Chezzetcook,* Cow Bay, Margaret's Bay, Hammond's Plains, *Montagu, Waverley, Oldham, Old Tangier* or *Mooseland, Tangier, Middle* and *Upper Musquodoboit,* Killag, Scraggy Lake, Mosher River, and other places, gold, etc., in quartz, mispickel abundant at Montagu.

Waverley.—Soda felspar in quartz.

Beaver Bank.—Slates for building and paving.

Tangier.—Graphic granite, tourmaline, peat bogs; four miles N., tin ore.

Musquodoboit.—Gypsum, *fireclay,* manganese, *plumbago,* titaniferous iron ore in schist, molybdenum.

Jeddore.—Wad. Ship Harbour.—Wad.

Caledonia.—Chaplin's Saw Mills, *chiastolite.* Nelson's, talcose schist.

COLCHESTER COUNTY. Five Islands.—East River, *barytes!* calcite, dolomite, [ankerite], hematite, copper pyrites, plumbago. Indian Point, malachite, magnetite, red copper, tetrahedrite.

Pinnacle Island.—*Analcime,* calcite, *chabazite!* natrolite, siliceous sinter. Moose Island, stilbite.

[A few miles from the shore, in slates, marble [*white!* and variegated.]

Londonderry.—Seat of Acadia Iron Mines. On branch of Great Village River, *barytes,* ankerite, hematite, limonite, magnetite; Cook's Brook, ankerite, hematite; Martin's Brook, hematite, limonite; east of Great Village R., on high ground, hematite, limonite; Folly R., below Falls, ankerite, iron pyrites, on high land east of river, ankerite, limonite, hematite; on Archibald's land, ankerite, barytes, hematite, limonite often as ochres worked for pigments; seven miles from outlet of Folly R., copper ore [small veins]; five miles from Folly Village, valuable quarry of building stone and grindstone.

Onslow.—Mountain, marble, [red and chocolate]; East Mountain, *pyrolusite*, manganite, gypsum, clays and marls, *umber*. Hoar's, wad. Tatamagouche, *freestone, grindstone*.

Kempt-town.—Salt-spring. Earltown, mineral spring?

Salmon River.—Earthy plumbago; south of, coal, copper pyrites, hematite.

Brookfield.—Two miles from station, *limonite* [in numerous boulders, some of huge dimensions], barytes.

Shubenacadie River;—Border of Hants county, *anhydrite, gypsum*, calcite, *barytes*, hematite, manganese; at the canal, iron pyrites; near railway station and for miles in vicinity, *clay beds* [brick yards, potteries].

Stewiacke.—*Gypsum, barytes*. Upper Stewiacke, gold.

Gay's River.—*Gold* in conglomerate, *argentiferous galena* in limestone, gypsum holding quartz sand.

Pictou County. Pictou.—*Freestone, clay beds* [potteries], jet, manganese; Roder's Hill, barytes; Carribou River, grey copper and malachite in lignite; West River, near Durham, copper ores in lignite; Little Harbour, grey marble.

New Glasgow.—Albion Coal Basin, *coal* [extensively worked by several companies], *oil-coal*, clay ironstone, hematite, *clay beds* [potteries].

East River.—Moulding sand; five miles from New Glasgow, limestone; Fishpools, copper ores in sandstone; 7 miles, gypsum and green copper ore; Springville, 11 miles, *specular iron ore, brown hematite*, manganese; Bridgeville, 12 miles, M'Gillivray's, *iron ores*; west side of E. River, *gypsum*; Elmsville, 14 miles, marble [green, containing pyrites, and drab with green streaks].

Fraser's Mountain.—*Marble* [grey, curiously waved].

Middle River.—Slates.

Sutherland's River.—Coal. Merigomish.—pencil stone [clay slate].

French River.—Gold.

Salt Springs.—Dense Brine. Gairloch.—Acid water?—New Lairg.—Iron ore.

Antigonish County. Antigonish.—*Gypsum*, alabaster, *fibrous gypsum*, vivianite, *limestone* [building stone]; *brine* [worked], near the town? *coal, oil-coal;* Lochaber Road, *brown ochre;* Braley's Brook, *copper pyrites in ochre;* Morristown, *epidote in trap;* Ogden's Lake, *gypsum;* Ballantyne's Cove, *gypsum;* Strait of Canseau, Tracadie, *limestone,* gypsum.

Lochaber and Polson's Lakes.—*Iron and copper pyrites,* brown hematite; Polson's Lake, garnets, iron pyrites [modified crystals, not *in situ*]. South River, *quartz* [large colourless crystals.]

Arisaig.—Galena with lignite, jasper [striped], *iron pyrites* [fine crystals] *copper ore, steatite.* Frenchman's Barn, barytes.

GUYSBOROUGH COUNTY. Guysborough.—Galena, hematite. *Country Harbour, Isaac's Harbour, Sherbrooke, Wine Harbour,* and other places not named, *gold,* etc., in quartz, and sometimes in washings. Milford River, Shore of Chedabucto Bay, gold.

Sherbrooke.—Anatase [fine but small crystals] in quartz, *peat bogs.* Cape Canseau.—Chiastolite in slate, granite, gneiss, mica slate, clay slate.

Port Mulgrave, Strait of Canseau.—*Gypsum* exported.

CAPE BRETON ISLAND.

RICHMOND COUNTY.—*Little River, Carribou,* Inhabitants Basin, coal. Little River, Arichat, St. Peter's, Lennox Passage, *limestone, gypsum;* River Inhabitants, *limestone.*

INVERNESS COUNTY.—Port Hood, Mabou, Broad Cove, Chimney Corner, *coal.*

Judique, Smith's Island, Mabou Harbour, Margarie, Plaster Cove, *limestone, gypsum;* Port Hood Island, Margarie, *freestone;* west of Plaster Cove, barytes, calcite; nearer the Cove, calcite, chalybite. fluor [blue, small crystals].

M'Kenzie's River.—Bitumen in calcite, apple green calcite, silver in nuggets and in sparry mineral veins, fluor spar [octahedral crystals more than an inch across], galena; Limbo Cove, talcose schist [abundant] full of garnets; Cheticamp, green and blue carbonate of copper, grey and yellow copper ore in calcite, chrysocolla, *red felspar!!* [perfect crystals, 8 inches long] in syenite; fourteen miles N. E. of Cheticamp, barytes; between this and Grand Anse, hematite; Grand Anse, sulphur spring; between Grand Anse and Cape St. Lawrence, magnetic iron; Bay St. Lawrence, coal? Jerome River, native copper, vitreous copper, poonah earth in trap.

Table lands, north east of county, *peat bogs.*

Between Strait of Canseau and Port Hood.—Long Point River and a second river, not named, gold; west of Cheticamp, Margarie R. and another, gold; east of Cheticamp, Steep Mountain or Little River, and Lazar or Red Point River, gold.

VICTORIA COUNTY.—Wagamatcook, or Middle River, *gold*, in quartz and washings, bismuth in nuggets, up to size of pigeon's eggs, with gold and titaniferous iron sand; Baddeck R., gold; Watchabuckt, Little Bras d'Or, gold and sulphuret of silver, iron and copper pyrites in quartz, argentiferous galena in quartz, gold in conglomerate; Cape North, *copper ore*, gold? syenite and porphyry, *peat bogs* on table lands in C. North district.

St. Ann's, North River.—Soap stone with white marble and quartz.

Whycocomagh.—*Freestone*, *marble*.

New Campbellton.—*Coal*.

CAPE BRETON COUNTY. Little Bras d'Or, Sydney, Lingan, Little Glace Bay, Big Glace Bay, Cow Bay, Mira Bay.—*Coal* [extensively worked by several companies]; Sydney district, clay iron stone in nodules.

St. Andrew's Channel, Sydney, Boulardarie Island, Mira Bay, etc., *limestone, gypsum, freestone, grindstone*.

East of Bras d'Or.—Syenite, porphyry. Near Bras d'Or.—*Marble, mineral springs*.

Gabarus Bay.—Molybdenum [crystals] in quartz.

In the Island of Cape Breton, no locality specified, topaz reported; also native sulphur.

ADDITIONAL IN N. S.

King's County Starr's Point.—Magnetic iron, gypsum, [selenite?], Iceland spar, [all in sandstone?].

Colchester County. Five Islands.—Manganese, umber.

CHAPTER XV.

NOTES ON THE RESERVATION OF MINERALS.

THE mineral reservations not being uniform throughout the province I thought it desirable to give some account of them and applied to W. A. Hendry, Esq., Deputy Commissioner of Crown Lands, who very kindly undertook to furnish the information required, and has favoured me with the statements in this chapter. These have not been obtained without considerable trouble and Mr. Hendry found that it would be an endless work to make a complete return of the mineral reserves, partly in consequence of the counties having been so often divided: for example, Digby and Annapolis once formed a single county; Shelburne included Yarmouth; Halifax included Pictou and Colchester. What he has furnished, however, is perhaps sufficient to give a general idea of the extent of land granted with the reservation only of the Royal Mines and the addition of lead, copper, and coal, in some instances. After the grant to the Duke of York in 1826 of all the mines and minerals of the province the reserves included every mineral substance whatever, but in 1858, when the arrangement was made by which the mines were surrendered to the Crown, an act was passed by the Legislature giving up to the grantees of land all the minerals previously reserved excepting gold, silver, tin, lead, copper, coal, iron, and precious stones, and, in the words of the act passed 28th March, 1858, "all other mines, minerals, ores, and earths, including ironstones, limestones, slate stone, slate rock, gypsum and clay," are now granted with lands. With regard to grants made in earlier years, those previous to 1759 have no reserves of minerals, with the exception of a single lot in 1752, where, perhaps, there may be an error in the date. Of the grants in 1759 some reserve no minerals; others reserve gold, silver, precious stones, and lapis lazuli.* From July 12th 1764, lead, copper, and coals were added to the reserved minerals in all grants of moderate size but nearly two million acres were granted in very large tracts (mostly in October 1765) to Alexander M'Nutt and others on certain conditions with respect to settling and cultivation. In these grants, gold, silver, and coals are the only minerals reserved. After the 4th of November 1766 the precious stones were omitted, and gold, silver, lead, copper, and coals only reserved. All mines and minerals with the exception of the reserves are expressly granted, consequently the proprietors of land held by grants antecedent to July 12th 1764 own all the minerals except gold, silver, precious stones, and lapis lazuli. Those who hold under the M'Nutt class of grants own all except gold, silver, and coals. Those who hold under common grants between the 12th July 1764 and November 4th 1766 own all except gold, silver, precious stones, lapis lazuli, lead, copper, and coals, and those whose grants are later than the date last given are the owners of all except gold, silver, lead, copper, and coals. It is provided in the act of 1858 that it shall apply to no mines or minerals which, at the time of its coming into operation, shall not, by virtue of the surrender or otherwise, be vested in the Crown, or be under the control of the Legislature of this province, nor to any mines or minerals which shall be subject to any grant, sale, lease, or disposition thereof, in force and subsisting at the time of its coming into operation, and that it shall not affect the then existing rights of any person or body corporate.

Most of the grants to the old townships were given to the committees of the persons who designed to settle them. These committees brought lists of the names of the intending settlers, returned to New England after they had received the grants, and brought back the settlers and their cattle. These first grants either reserve no minerals, or else gold, silver, precious stones, and lapis lazuli, but the greater part of them were resigned and new ones procured, the old grants being stated to be insecure; many persons, however, procured separate grants for shares in these townships during the period that elapsed between the first and second grants. It

*Lapis lazuli is a mineral found in granite, syenite, and crystalline limestone. The richly coloured varieties are highly esteemed for various ornamental purposes but the value of these has diminished of late years. The expensive pigment ultramarine was not long ago prepared exclusively from lapis lazuli but it is now made artificially of quality equal to that from the mineral at about one two hundredth part of the cost. Lapis lazuli has not been found in the province so far as I am aware.

does not appear certain that these persons gave up their rights when the township grants were surrendered, if they did not they would, in most cases, hold a right to all minerals except gold, silver, and precious stones.

In each county there remain large tracts of ungranted land: the following are the most important grants of earlier years.

ANNAPOLIS COUNTY.

	Acres
Annapolis township, granted in 1765, with reservation of gold, silver, lead, copper, coal, precious stones, and lapis lazuli	65,000
Granville township, reservations as before	52,600
Wilmot township, granted at different times, some of the grants reserve no minerals, while others reserve gold, silver, lead, copper, coals, and precious stones; about	56,000
	173,600

ANTIGONISH AND GUYSBORO' COUNTIES.

Hollowell grant, reserved, gold, silver, and coal	20,000
Byron ,, ,, ,, ,, ,, ,,	10,000
Montgomery,, ,, ,, ,, ,, ,,	20,000
Ad'ml. Lord Colville grant, reserved, gold, silver, and coal	20,000
Milford Bay ,, ,, ,, ,, ,, ,,	20,000
	90,000

Several smaller grants reserve lead, copper, coal, gold, silver, and precious stones: to sum up, these two counties stand thus:—

Grant reserving gold, silver, and coal	250,000
,, ,, gold, silver, lead, copper, and coal	248
,, ,, gold, silver, lead, copper, coal, precious stones, and lapis lazuli	8,778
	259,026

COLCHESTER COUNTY.

Onslow township, granted 21st Feb., 1759, with reservation of gold, silver, lead, copper, and coal	50,000
Tatamagouche, Desbarres grant, 25th Aug. 1765, reserved, gold, silver, and coal	20,000
Londonderry township, grant 30th Oct., 1765, reserved, gold, silver, precious stones, lead, copper, and coal	100,000
Truro Township, grant 31st Oct., 1765, reserved, gold, silver, precious stones, lead, copper, and coal	50,000
Upper Stewiacke, grant 29th July, 1769, reserved, gold, silver, and coals	100,000
	320,00

CUMBERLAND COUNTY.

	Acres.
Cumberland grant, 12th Oct., 1763, reserved, gold, silver, precious stones, and lapis lazuli	75,000
Amherst, reserved, gold, silver, and coal	100,000
	175,000

DIGBY COUNTY.

St. Mary's Bay grant, 31st Oct., 1765, reserved, gold, silver, and coal	125,000

HALIFAX COUNTY.

No large township grants: the grants rarely exceeded three or four thousand acres and they may be classed as follows:—

	Acres.
Between 1752 and 1754, without any mineral reserves	32,909
,, 1759 and 1764, reserved, gold, silver, precious stones, and lapis lazuli	71,366
,, 1764 and 1766, reserved, gold, silver, lead, copper, coal, precious stones and lapis lazuli	84,180
,, 1766 and 1782, reserved, gold, silver, lead copper, coal	116,721
,, 1765 and 1773	468,000
	773,176

HANTS COUNTY.

Falmouth grant, 11th June, 1761, reserved, gold silver, precious stones, and lapis lazuli	50,000
Newport grant, reserves gold, silver, precious stones	58,000
Other small grants, reserving gold, silver, precious stones and lapis lazuli, making a total of	129,366
	227,366

KINGS COUNTY.

Horton, reserved, gold, silver, precs. stones, & lapis lazuli	100,000
Cornwallis, ,, ,, ,, ,, ,, ,, ,,	100,000
Between 1764 and 1866 grants reserving gold, silver, lead, copper, precious stones, and lapis lazuli	58,857
	258,857

QUEENS COUNTY.

Liverpool township, reserved, gold, silver, precious stones, lapis lazuli	100,000
North-west side of Liverpool, reserved, gold, silver, coal	200,000
South-west ,, ,, ,, ,, ,, ,,	100,000
	400,000

RESERVATION OF MINERALS. 215

LUNENBURG COUNTY.

	Acres.
Chester township, reserved, gold, silver, precious stones, lapis lazuli............................	100,000
New Dublin, reserved, gold, silver, precious stones, lapis lazuli................................	200,000
Between 1752 and 1758, with no mineral reserves.........	4,120
Between 1759 and 1766, reserved, gold, silver, precious stones, and lapis lazuli........................	71,049
Between 1766 and 1782, reserved gold, silver, lead, copper, coal	40,429
	415,597

PICTOU COUNTY.

Between 1765 and 1766, reserved, gold, silver and coal....	440,000
,, ,, ,, ,, ,, ,, lead, copper, coal...	10,000
	450,000

SHELBURNE COUNTY.

Barrington township, no minerals reserved.............. 46,750

YARMOUTH COUNTY.

Between 1760 and 1764, reserved, gold, silver, precious stones, lapis lazuli............	2,500
,, 1764 and 1766, reserved, gold, silver, precious sts., lap. laz., lead, copper, coal	21,000
,, 1766 and 1782, reserved, gold, silver, lead, copper, coal...'...............	152,548
	176,548

ISLAND OF CAPE BRETON.

The grants were not issued until 1786.

APPENDIX.

After the foregoing report was presented I received:

Report on the Gold Regions of Nova Scotia, by Dr. Sterry Hunt, F.R.S., addressed to Sir W. E. Logan, F.R.S., Director of the Geological Survey of Canada.

GOLD PRODUCT OF THE PROVINCE.

The amount of gold officially reported from Jan. 1st to Sep. 30th, 1868, was 15,459 oz. 6 dwt. 17 grs., making the total product so given at the latter date 140,712 oz. 8 dwt. 20 grs. There have since been receipts of bar gold in Halifax to the amount of 2,467 oz. up to Dec. 3rd, 1868, giving in round numbers, as the partial yield up to latter date, the total of 143,179 ounces.

COPPER ORE OF CAPE NORTH, CAPE BRETON.

A vein has lately been reported, as traced for six miles in this district, consisting of eight feet thickness of gangue containing copper ore scattered throughout, with a centre of eighteen inches composed of good yellow ore.

PROVINCIAL SALT AT THE LATE EXHIBITION.

EXHIBITS.

No. 1. Nova Scotia Salt Works—Prize awarded.
No. 2. J. H. Hewson, Pictou.

EXPORT OF MINERALS FROM WINDSOR IN 1868.

	Tons of 2,240 lb.	Value.
Gypsum up to Dec. 12th	52,310	$47,079
Moulding Sand	290	290
Manganese	12	300

MARBLE IN CAPE BRETON.

A recent article (Dec. 7th) in the P. E. I *Summerside Progress* states that large deposits of marble have been found nearer Bras d'Or Lake than those mentioned in the report. Mr. Brown, of St. Eleanor's, P. E. I., has secured quarrying rights over about 1600 acres containing beds of marble of 7 or 8 varieties, (black, white, veined, spotted, and of rare flesh colour; some of superior quality), practically inexhaustible, and within a few rods of deep water. The marble hill, called North Mountain, is on the N. shore of the lake, or West Bay, 25 miles E. of Ship Harbour, and in sight of St. Peter's canal. Mr. Brown is pushing forward such operations as will enable him to begin shipping marble in the coming season.

COAL MINES.

No official returns relative to coal have been made since those referred to in the Report. Important operations have been in progress and it is fully expected that next spring will see increased activity in the development of the coal mines.

Coal of Drummond Colliery, Bear Creek, Pictou Co.—Operations were begun here in November, 1867, and since that time a double shaft has been sunk on the seam (Report, table p. 10) to a depth of 730 feet, with lateral galleries at intervals of from 200 to 300 feet, and much coal has been shipped to New York, Boston, Portland, Quebec, and Montreal, at which places it can be laid down cheaper than Scotch coal. The coal has been tested by the manager of the Pictou Gas Works who certifies that the yield, 10,000 cubic feet of gas to the ton, is greater than that of the best coals of the district, (compare table p. 33 of Report), and also, that the coke is more per ton and of better quality. It appears to give 1500 cubic feet more gas per ton than the average coal from Newcastle.

www.ingramcontent.com/pod-product-compliance
Lightning Source LLC
Chambersburg PA
CBHW031818230426
43669CB00009B/1182